Outer Space—
Battlefield of the Future?

sipri

Stockholm International Peace Research Institute

SIPRI is an independent institute for research into problems of peace and conflict, especially those of disarmament and arms regulation. It was established in 1966 to commemorate Sweden's 150 years of unbroken peace.

The institute is financed by the Swedish Parliament. The staff, the Governing Board and the Scientific Council are international. As a consultative body, the Scientific Council is not responsible for the views expressed in the publications of the Institute.

Governing Board

sipri

Stockholm International Peace Research Institute
Sveavägen 166, S-113 46 Stockholm, Sweden
Cable: Peaceresearch, Stockholm
Telephone: 08-15 09 40

Outer Space—
Battlefield of the Future?

sipri

Stockholm International Peace Research Institute

Taylor & Francis Ltd   Crane, Russak & Company, Inc.
London   New York
1978

First published 1978 by Taylor & Francis Ltd., London
and Crane, Russak & Company, Inc., New York

Copyright © 1978 by SIPRI
Sveavägen 166, S-113 46 Stockholm

ISBN 0 8448 1312-5

Library of Congress Catalog Card Number 77-26368

Printed and bound in the United Kingdom by
Taylor & Francis (Printers) Ltd, Rankine Road,
Basingstoke, Hampshire RG24 0PR

Preface

Publicity about satellites, and about space activities in general, normally focuses on their peaceful applications. Consequently, there is little public debate about the military use of space, in spite of the fact that about 60 per cent of US and Soviet satellites are military ones. Since the space age began, 1386 military satellites are known to have been launched by end-1976 – 563 by the USA, 899 by the USSR, 5 by the USA for the UK, 3 by the USA for France and 5 by France itself, 2 by China, and 4 by the USA for NATO.

Satellites for military communications over both short and long distances, for weather prediction, and for accurate navigation are among the types used by the military. Other important types are photographic and electronic reconnaissance satellites to identify all sorts of military targets, satellites to position military targets accurately, satellites to give early warning of the launching of enemy missiles, and those capable of intercepting and destroying orbiting enemy satellites.

Up to the end of 1976 the USA had spent about $30 000 million on its military space activities, about one-third of the total sum spent on space. The cost of the Soviet military space programme is kept secret, but the magnitude of the effort is similar to that of the USA.

China, France, the UK and NATO also operate military satellites. The exact purpose of the Chinese satellites is not known, but photographic reconnaissance is probably included.

Considerable efforts are currently being made to increase the survivability in war of military satellites. Research into, for example, protection of orbiting satellites against nuclear attack from a hostile satellite is actively under way. Also under investigation are detection systems for early warning of satellite attacks, based on the surveillance of space by ground- and space-based sensors.

In the long term, the most revolutionary military technological development may turn out to be the use of navigational and geodetic satellites to guide missiles on to their targets. There is no reason why CEPs of a few metres over intercontinental ranges should not be obtained by these means. Moreover, the space-based navigation system of one country may be used by others for military purposes. This, coupled with the almost inevitable proliferation of, for example, cruise missile technology, is an extremely worrying prospect.

The other side of the coin is the useful role of satellites in verifying, by 'national technical means', some arms control agreements.

For these reasons, availability of information about the military use of space is important. We hope that this book will provide such information for those working for the control of military activity in space.

This book was written by Dr. Bhupendra M. Jasani, a research fellow at SIPRI.

December 1977

Frank Barnaby
Director

Contents

Abbreviations, acronyms and conventions XV

Chapter 1. Introduction 1

Chapter 2. Some basic concepts of orbital characteristics 4
    I. Orbital dynamics 5
    II. Orbital perturbations 9
        Effects of the Earth's rotation – Relativistic effects – Effects due to
        the asymmetry of the Earth – Gravitational effects of the Sun and
        Moon – Effects due to the Earth's atmosphere – Effects of magnetic
        drag – Effects of solar-radiation pressure

Chapter 3. Reconnaissance satellites 12
    I. Photographic reconnaissance satellites 12
        Satellite ground tracks – Satellite orbits – Space photography –
        The US programme – The Soviet programme – The Chinese pro-
        gramme – The French programme
    II. Electronic reconnaissance satellites 39
        Satellite orbits – The US programme – The Soviet programme
    III. Ocean-surveillance satellites 43
        The US programme – The Soviet programme
    IV. Early-warning satellites 44
        Satellite orbits – The US programme – The Soviet programme
    V. Nuclear-explosion detection satellites 47
        The US programme – The Soviet programme
Appendix 3A. Tables of photographic reconnaissance satellites 49
Appendix 3B. Tables of electronic reconnaissance satellites 82
Appendix 3C. Tables of ocean-surveillance satellites 91
Appendix 3D. Tables of early-warning satellites 93

Chapter 4. Communications satellites 97
    I. Satellite orbits 98
    II. Satellite transponder characteristics 102
    III. The US programme 105
    IV. The Soviet programme 107
    V. The British programme 110
    VI. The NATO programme 111
    VII. The French programme 112
Appendix 4A. Tables of communications satellites 114

Chapter 5. Navigation satellites 131
   I. Some basic concepts.. 131
  II. The US programme 135
 III. The Soviet programme..................................... 137
Appendix 5A. Tables of navigation satellites 139

Chapter 6. Meteorological satellites............................. 144
   I. Satellite orbits 144
  II. The US programme 145
 III. The Soviet programme..................................... 147
 IV. The British and French programmes......................... 149
Appendix 6A. Tables of meteorological satellites 150

Chapter 7. Geodetic satellites 158
   I. Satellite orbits 160
  II. The US programme 160
 III. The Soviet programme..................................... 162
 IV. The French programme 162
Appendix 7A. Tables of geodetic satellites 163

Chapter 8. Interceptor/destructor satellites and FOBSs 167
   I. Some anti-satellite systems 167
     The US programme – The Soviet programme
  II. Satellite tracking systems and facilities 176
     US facilities – Soviet facilities
 III. Fractional orbital bombardment systems (FOBSs) 179
Appendix 8A. Tables of interceptor/destructor satellites and FOBSs 181

Chapter 9. Conclusions 184

References 189

Glossary 194

Index 199

Tables and Figures

Chapter 1. Introduction

Figure
1.1.   Recovery of a satellite in orbit by a space shuttle vehicle   1

Chapter 2. Some basic concepts of orbital characteristics

Figures
2.1.   Stages between the time a satellite is lifted off the ground and when it starts to orbit the Earth   4
2.2.   Satellite paths resulting from injection angles and velocities of greater and less than V_s   6
2.3.   Various types of orbits used for communications satellites   6
2.4.   The geometric terms used in satellite orbits – showing the six basic parameters   7
2.5.   The Earth's orbit showing the directions of its revolution round the Sun and its rotation about its axis   8
2.6.   Some satellite orbital parameters   8
2.7.   Mathematical treatment as harmonics of anomalies in the Earth's gravitational field, resulting from the slightly pear-shape of the Earth   10

Chapter 3. Reconnaissance satellites

Table
3.1.   US photographic reconnaissance satellites, launchers and orbital inclinations   30

Figures
3.1.   The electromagnetic spectrum   12
3.2.   Ground tracks over India of the US 1974-20A Big Bird satellite, launched on 10 April 1974 at an orbital inclination of 94.52°, 16–25 May   15

3.3. Ground tracks over India of the Soviet Cosmos 653 satellite, launched on 15 May 1974 at an orbital inclination of 62.81°, 16–25 May   16

3.4. Ground tracks over Cyprus, Greece and Turkey of the Soviet Cosmos 666 satellite, launched on 12 July 1974 at an orbital inclination of 62.81°, 13–25 July   18

3.5. Ground tracks over Cyprus, Greece and Turkey of the US 1974-65A satellite, launched on 12 July 1974 at an orbital inclination of 110.51°, 15–30 August   19

3.6. Ground tracks over Cyprus, Greece and Turkey of the Soviet Cosmos 667 satellite launched on 25 July 1974 at an orbital inclination of 64.98°, 25 July–7 August   20

3.7. Ground tracks over Cyprus, Greece and Turkey of the US 1974-20A satellite, launched on 10 April 1974 at an orbital inclination of 94.52°, 14–28 July   21

3.8. Simple geometry of the satellite camera system   22

3.9. Ground resolution as a function of focal length (satellite altitude = 150 km)   23

3.10. The Chicago area photographed from the Skylab space station, from an altitude of about 440 km   24

3.11. An enlarged section of a photograph taken from the Skylab satellite showing the MacDill Air Force Base, Florida   25

3.12. A US Discoverer military photographic reconnaissance satellite. .   26

3.13. The nose cone of a Discoverer satellite similar to one used in recovering a capsule from orbit round the Earth   27

3.14. Vandenberg Air Force Base (WTR), showing Launch complex areas.   28

3.15. A Discoverer satellite capsule   31

3.16. Re-entry of the Capsule from a US reconnaissance satellite   31

3.17. A Titan-3D rocket launching a Big Bird reconnaissance satellite into orbit   32

3.18. A full-scale model of Cosmos 381, displayed at the 1971 Paris Air Show   33

3.19. Photograph of the Tyuratum Cosmodrome in the Soviet Union, taken from a US Landsat satellite   34

3.20. Drawing made from a photograph taken from a Landsat satellite of the Soviet northern launch centre   35

3.21. Ground tracks of a Soviet electronic reconnaissance satellite, Cosmos 749 (1975-62A) for a 14-day period   42

Appendix 3A. Tables of photographic reconnaissance satellites

Tables
3A.1. US photographic reconnaissance satellites 49
3A.2. Soviet photographic reconnaissance satellites 60
3A.3. Possible Chinese photographic reconnaissance satellites 81

Appendix 3B. Tables of electronic reconnaissance satellites

Tables
3B.1. US electronic reconnaissance satellites 82
3B.2. Possible Soviet electronic reconnaissance satellites 86

Appendix 3C. Tables of ocean-surveillance satellites

Tables
3C.1. US ocean-surveillance satellites 91
3C.2. Possible Soviet ocean-surveillance satellites 91

Appendix 3D. Tables of early-warning satellites

Tables
3D.1. US MIDAS, Vela and other early-warning satellites... 93
3D.2. Possible Soviet early-warning satellites 96

Chapter 4. Communications satellites

Figures
4.1.   Satellite periods for various altitudes 98
4.2.   Satellite coverage geometry 99

4.3. Changes of percentage of Earth's surface visible from different satellite altitudes.. 100
4.4. Artist's impression of a US DSCS II communications satellite.... 106
4.5. The Soviet communications satellite Molniya 1................ 108
4.6. The British communications satellite Skynet 2B, shown mated to the three-stage Delta vehicle................................ 110
4.7. The NATO communications satellite NATO 3B, being readied for launch 112
4.8. The French–West German communications satellite Symphonie.. 113

Appendix 4A. Tables of communications satellites

Tables
4A.1. US communications satellites 114
4A.2. Possible Soviet communications satellites 120
4A.3. British communications satellites with possible military applications 129
4A.4. NATO communications satellites 130
4A.5. French communications satellites with possible military applications 130

Chapter 5. Navigation satellites

Figures
5.1. The Doppler principal applied to a satellite................... 132
5.2. The US navigation satellite Transit IIA 135

Appendix 5A. Tables of navigation satellites

Tables
5A.1. US navigation satellites.................................... 139
5A.2. Soviet navigation satellites 142

Chapter 6. Meteorological satellites

Figures

6.1.   A US defence meteorological satellite for use by the USAF   145
6.2.   Photograph of the weathern pattern on the Earth's disc, taken from the first US synchronous meteorological satellite SMS-1 on 28 May 1974, with a resolution of 0.9 km   146
6.3.   The Soviet meteorological satellite Cosmos 144   148

Appendix 6A. Tables of meteorological satellites

Tables

6A.1. US meteorological satellites   150
6A.2. Soviet meteorological satellites   154
6A.3. British meteorological satellites with possible military applications   157
6A.4. French meteorological satellites with possible military applications   157

Chapter 7. Geodetic satellites

Figures

7.1.   Methods of connecting points on land masses using satellites   159
7.2.   Photograph of the first successful operational test of the US geodetic satellite ANNA's flashing light shown within the five circles . . . .

Appendix 7A. Tables of geodetic satellites

Tables

7A.1. US geodetic satellites   163
7A.2. Possible Soviet geodetic satellites   165
7A.3. French geodetic satellites with possible military applications   166

Chapter 8. Interceptor/destructor satellites and FOBSs

Table
8.1.   Effects of atmospheric conditions on various lasers   170

Figure
8.1.   Radio signal tracking by measuring satellite beacon signals   177

Appendix 8A. Tables of interceptor/destructor satellites and FOBS

Tables
8A.1. Possible Soviet interceptor/destructor satellites   181
8A.2. Soviet fractional-orbital bombardment systems   182

Chapter 9. Conclusions

Table
9.1.   Summary of possible military satellites, by type of mission   186

Abbreviations, Acronyms and Conventions

Abbreviations

cm	centimetre
deg	degree
GHz	gigahertz
GMT	Greenwich Mean Time
h	hour
kg	kilogram
km	kilometre
kW	kilowatt
m	metre
MeV	megaelectronvolt
MHz	megahertz
Mt	megaton
MW	megawatt
min	minute
mm	millimetre
mn	million
s	second
K	Kelvin
°C	degree Celsius

Acronyms

A-A	Agena-A launcher
A-B	Agena-B launcher
A-D	Agena-D launcher
ABM	Anti-Ballistic Missile
AFSATCOM	Air Force Satellite Communication
AMS	Advanced Meteorological Satellite
ANNA	Army, Navy, NASA, Air Force
ARPA	Advanced Research Projects Agency
ATS	Applications Technology Satellite

BMEWS	Ballistic Missile Early-Warning System
CNES	Centre National d'Etudes Spatiale
COAT	Coherent Optical Adaptive Technique
Comsat	Communications Satellite Corporation
CSC	Communications Satellite Corporation
DATS	Daspun Antenna Test Satellite
DMSP	Data Meteorological Satellite Program
DoD	Department of Defense
DSCS	Defense Satellite Communications System
DTEN	Direction Technique des Constructions et Armes Navales
ECCM	Electronic Counter-Counter Measure
ECM	Electronic Counter-Measure
Elint	Electronic Intelligence
ESSA	Environmental Science Service Administration
ETR	Eastern Test Range (Cape Kennedy, Florida)
FLTSATCOM	Fleet Satellite Communication
FOBS	Fractional Orbital Bombardment System
GEOS	Geodynamic Experimental Ocean Service Administration
GOES	Geostationary Operational Environmental Satellite
ICBM	Intercontinental Ballistic Missile
IDCSP	Initial Defense Communication Satellite Program
IDSCS	Initial Defense Satellite Communication System
IMEWS	Integrated Missile Early-Warning System
Intelsat	International Telecommunications Satellite Consortium
IRBM	Intermediate Range Ballistic Missile
ITOS	Improved TIROS Operational Satellite
Lageos	Laser Geodynamic Satellite
Landsat	Land Satellite (Earth Resources Technology Satellite)
LASP	Low Altitude Surveillance Platform
LES	Lincoln Experimental Satellite
LTTAT	Long-Tank Thrust Augmented Thor Launcher
Marisat	Marine Communications Satellite
MIDAS	Missile Defense Alarm System
NASA	National Aeronautics and Space Administration
NATO	North Atlantic Treaty Organization
NNSS	Navy Navigation Satellite System
NOAA	National Oceans and Atmosphere Administration

NORAD	North American Air Defense Command
NOSSI	Navy Ocean Surveillance Satellite
NTS	Navigation Technology Satellite
OTH	Over-the-Horizon (radar)
OV	Orbiting vehicle
Pageos	Passive Geodetic Satellite
PDM	Pulse Duration Modulation
PMTC	Pacific Missile Test Center
PTB	Partial Test Ban Treaty
RAE	Royal Aircraft Establishment
RBV	Return-Beam-Vidicon (camera)
RCA	Radio Corporation of America
R&D	Research and Development
RF	Radio Frequency
RMU	Remote Manoeuvring Unit
SAINT	Satellite Inspector Technique
SALT	Strategic Arms Limitation Talks
SAMOS	Satellite Missile Observation System
SAMSO	Space and Missiles Systems Organization
SAMTEC	Space and Missile Test Center
Satcom	Satellite Communication
SCORE	Signal Communication by Orbiting Relay Equipment
SDS	Satellite Data System
SHAPE	Supreme Headquarters, Allied Powers, Europe
SHF	Super-High Frequency
SLBM	Submarine-Launched Ballistic Missile
SMS	Synchronous Meteorological Satellite
SPADATS	Space Detection and Tracking System
SPASUR	Space Surveillance System
SR	Solrad
Tacsat	Tactical Satellite
Tacsatcom	Tactical Satellite Communications
TAT	Thrust Augmented Thor launcher
Th	Thor launcher
TIMATION	Time Navigation
TIP	Transit Improvement Program
TRANET	Transit Network

UHF	Ultra High Frequency
USAF	US Air Force
USN	US Navy
WSMR	White Sands Missile Range
WTR	Western Test Range (Vandenberg AFB, California)
WWMCCS	World-Wide Military Command and Control System
YAG	Yttrium–Alumina–Garnet

Conventions

..	Information not available
—	None
?	Uncertainty about the satellite designation or other data
<	Less than the number given
>	More than the number given
⩽	Equal to or less than the number given
⩾	Equal to or more than the number given
[1]	Square-bracketed numbers in the chapters refer to the list of references at the end of the book

1. Introduction

In November 1976 a Soviet satellite was launched into orbit from Plesetsk. Four days later a similar cylindrical satellite was launched from Tyuratam, manoeuvred close to the first satellite and then moved away again. Previous tests had shown that the manoeuvrable satellite can be exploded on command from an Earth station. It takes little imagination to comprehend what use could be made of a satellite which can approach another and explode near it.

In early 1977 taxi trials of a space shuttle orbiter mounted atop a US Boeing 747 aircraft were successfully completed. The manoeuvrable space shuttle, equipped with massive 60-ft doors which can open and shut to permit deployment and recovery of objects in space, will be capable not only of inspecting an enemy satellite in flight but also of capturing and retrieving it (see Figure 1.1).

These recent spectacular developments in space technology of great sig-

Figure 1.1. Recovery of a satellite in orbit by a space shuttle vehicle

1

nificance to the military in both the USA and USSR are only the most extra-ordinary examples of what man is preparing for space. Other less spectacular developments have been made ever since the space age began – all potentially in preparation for the new battleground of outer space. Before 1957, when the first artificial Earth satellite was launched, the very idea of a satellite which could hunt down another and destroy it by small explosives or laser beams belonged to the realm of science fiction. In the two decades since then, man has left his planet and walked on the Moon, and has retrieved spectacular photo-graphs from the distant planets of Mars, Jupiter and Mercury. All these achieve-ments have distracted attention from the other remarkable but frightening military achievements in space at the same time as we have become used to the idea of having activity nearly beyond our comprehension going on in our own atmosphere and beyond. The prospects of what could happen and what might be allowed to happen must be examined before the most absurd and destructive type of science fiction we know becomes a reality for us within the coming decade.

It was believed in 1967 that outer space would for the foreseeable future be a 'zone of peace' after the ratification of the Outer Space Treaty by the USA and the USSR, as well as by 72 others by end-1976. But this illusion has been shattered by the past decade of revolutionary advances in military space tech-nology and by the realization that this treaty only prohibits 'the placing in orbit around the Earth of any objects carrying nuclear weapons or any other kinds of weapons of mass destruction'. In fact, in recent years about 60 per cent of both the US and the Soviet satellites launched into orbit have been military satellites. Although most of the space programmes are conducted by civilian agencies and are ostensibly for 'peaceful purposes', in fact the far greater portion of satellite programmes are in pursuit of military objectives and in preparation for carrying out military missions.

One type of satellite – reconnaissance satellites – has been thought to have a particularly direct role to play in arms control and disarmament endeavours. When the SALT I agreements were signed in 1972, the practice of using one's own 'national technical means' of verification of compliance with arms control treaties legalized the ongoing practice of using reconnaissance satellites for this purpose. Thus the USSR, for the first time, accepted the principle of 'open skies', proposed by the USA as long ago as July 1955, but only at satellite altitudes. Until China launched its first such satellite, the USA and the USSR were the only two nations with the reconnaissance capability for inspection of foreign territory.

In addition to this most well-known type of satellite, a great variety of other types have been launched by the two greater powers. For example, they both employ so-called early-warning satellites for advance warning of the launching of enemy missiles. Considerable developments are continuously being made in navigation, communications and meteorological satellites – develop-ments which are of use not least to the military in these countries.

For obvious reasons, a considerable degree of secrecy shrouds the satellite programmes of these three powers. However, it is possible to learn the launch dates and the basic characteristics of all the satellites launched†. Together with the estimated shape, weight and dimensions of each object and details such as which objects are payloads, which are spent rocket casings and which are debris, it is possible to learn or be able to conjecture the purpose or the mission for which the particular satellite was launched. This type of information is supplied by the Royal Aircraft Establishment (RAE) at Farnborough, UK. Much of the information on the basic characteristics is available from private observers; for example, for the Soviet Cosmos satellites the main source is a group associated with the Kettering Boys School in the UK. The Cosmos series covers a variety of missions and it is only through the study of repetitive patterns in orbit, the kind of debris associated with the flights, the types of signals they transmit and the timing of the satellite launches that it has been possible to classify most of the individual satellites by their various missions. The tables of satellites and their basic characteristics in the appendices of this book have been prepared on the basis of this type of information.

In order to appreciate the purpose for which a satellite is launched, it is necessary to understand the characteristics of its orbit (Chapter 2), from which can be derived the satellite's capabilities and suitability for performing various missions. This is followed by discussions of each of the satellite types: reconnaissance satellites (Chapter 3), communications satellites (Chapter 4), navigation satellites (Chapter 5), meteorological satellites (Chapter 6), geodetic satellites (Chapter 7) and interceptor/destructor satellites and FOBSs (Chapter 8). Sub-sections examine the relevant space programmes of the countries which launch such satellites. In the concluding chapter (Chapter 9), the capability and effectiveness of satellites for verifying the implementation of arms control agreements are discussed.

---

† Some data on all satellites launched are to be found in *Table of Earth Satellites*, published by the Royal Aircraft Establishment at Farnborough, UK. Detailed orbital parameters of all the satellites are also distributed by the NASA Goddard Space Flight Center. In future, such data may become generally available as a result of the recommendation on 26 November 1974 by the UN General Assembly of a text of the Convention on Registration of Objects Launched into Outer Space. According to the convention, when an object is launched into Earth orbit or beyond, the state which launches or procures the launching of the object should enter it in an appropriate register maintained by the state. The UN Secretary-General will also keep a central register in which the information given by the launching states will be recorded. The following information should be recorded: (a) the launching state or states; (b) an appropriate designation of the space object or its registration number; (c) the date and territory or location of launch; (d) basic orbital parameters, including nodal period, inclination, and apogee and perigee heights; and (e) general function of the space object. Such a register has been in existence since early 1960 and the data have been recorded on a voluntary basis, but the convention formalizes it on a mandatory basis.

2. Some basic concepts of orbital characteristics

A spacecraft follows an elliptical path if no other force than the Earth's gravity acts upon it. If the elliptical trajectory is not intercepted by the Earth's surface, the spacecraft will travel round the Earth and if it makes more than one circuit of the Earth without using thrust to counteract the pull of gravity, the spacecraft becomes a satellite. The orbit of a satellite is not a smooth elliptical path. The shape of the orbit is mainly influenced by the Earth's gravitational force but air drag, the pressure of sunlight and irregularities in the shape of the Earth continuously distort the orbit.

Satellites are usually placed in orbit by large multi-stage rockets which apply accelerating forces over a period of several minutes. Initially, a booster rocket spurting out large flames from its burning chemicals lifts a satellite off its launch pad. After a minute or so of vertical climb through the lower atmosphere, the launch vehicle is commanded from the ground station to pitch over into orbit (Figure 2.1), usually followed by several thrust-and-coast periods. During the latter periods, spent rocket stages are discarded. When the desired altitude has been reached, the final launch-vehicle stage injects the satellite into orbit with the required velocity and angle. When an orbit is confirmed from the worldwide tracking stations, and on command from the ground station, the

Figure 2.1. Stages between the time a satellite is lifted off the ground and when it starts to orbit the Earth

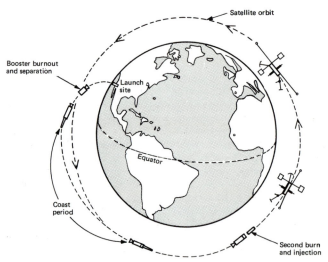

satellite becomes alive and begins to erect instrument booms, solar panels and telemetry antennas. The satellite orbits round the Earth, weaving sideways and up and down due to the electric and magnetic fields, radiation pressures, micrometeoroid impacts, the gravitational attraction of the Sun and the Moon, and the irregular gravitational forces of the Earth.

This chapter gives a short description of both orbital dynamics and perturbations.

1. Orbital dynamics

Assuming that the Earth is spherical with a spherically symmetric gravitational field and disregarding the gravitational effects of the Sun and the Moon on the satellite, the trajectory of a satellite is planar and either circular or elliptical with the Earth's centre of gravity as one of the foci of the ellipse. In the case of a circular orbit, the centrifugal force F_c counterbalances the gravitational force F_g so that

$$F_c = F_g \tag{2.1}$$

or

$$mV_s^2/R = GmM/R^2.$$

Therefore,

$$V_s = \sqrt{(GM/R)},$$

where

$m =$ the satellite mass (kg),
$R =$ the distance of the satellite from the centre of the Earth (m),
$V_s =$ the satellite velocity (m/s),
$M =$ the mass of the Earth (kg),

and

$G =$ the universal gravitational constant (N m^2/kg).

If the injection velocity is less than V_s, the spacecraft will return to Earth along an elliptical path (Figure 2.2). If, however, the horizontal injection velocity is greater than V_s, the orbit will be elliptical with the perigee at the injection point.

In the case of a circular orbit (most US communications satellites are launched into circular orbits) the period τ of a satellite is given by

$$\tau = 2\pi\sqrt{[(R_e + h)^3/\mu]} = 2.7644 \times 10^{-6} (6371.315 + h)^{3/2} \tag{2.2}$$

where R_e is the radius of the Earth in kilometres, h is the height of the satellite above the Earth in kilometres, μ is the Earth's gravitational constant (3.9862×10^5 km^3/s^2) and τ is in hours. If h is 35 900 km, the value of τ is the same

5

Figure 2.2. Satellite paths resulting from injection angles and velocities of greater and less than V_s

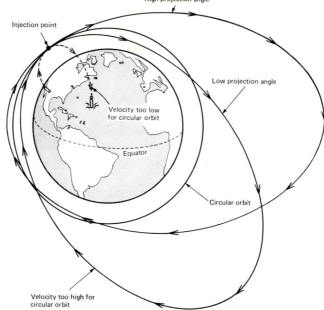

High projection angle

Injection point

Low projection angle

Velocity too low
for circular orbit

Equator

Circular orbit

Velocity too high for
circular orbit

Figure 2.3. Various types of orbits used for communications satellites

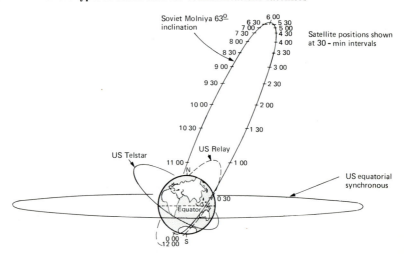

Soviet Molniya 63° inclination

Satellite positions shown
at 30 - min intervals

US Telstar

US Relay

US equatorial
synchronous

Equator

as the Earth's period of rotation about its own axis. Furthermore, if the orbital inclination is zero, the so-called synchronous satellite (Figure 2.3) will appear stationary with respect to the Earth. Such an orbit is achieved in several stages. First the satellite is launched into an elliptical transfer orbit. During the next stage, that of inclination adjustment, the satellite must be positioned for an apogee motor firing which injects it into a near-circular drift orbit. The satellite then drifts towards its final position above the equator.

Figure 2.4. The geometric terms used in satellite orbits – showing the six basic parameters
The Earth is at one of the foci of the ellipse.

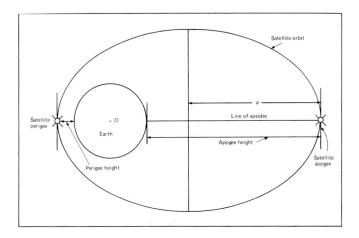

In order to understand what happens in practice, the above simple notion of circular orbit equating the centrifugal and the gravitational forces must be replaced by a more general conic section, the ellipse. Figure 2.4 shows such an orbit with the centre of the Earth at one of the foci of the ellipse. Six basic parameters, called the orbital elements, define such an orbit in space. Three of these describe the orbit itself and the remaining three define the position of the orbital plane relative to the Earth. The position of the satellite in orbit is not specified by the orbital elements.

The size and shape of the ellipse are determined by its semi-major axis a and its eccentricity e. These are two of the six orbital elements. During the lifetime of a satellite, most of the orbital elements are continuously changing, so that the values of all the elements must be given for a particular time. Often this time is chosen to be the passage time of the satellite through the ascending node (see below). The passage time T of the satellite at the perigee is the third orbital element and, together with a and e, defines the ellipse in a plane.

Man has taken considerable time to convince himself that the Earth is not the centre round which the Sun and other planets move. It is now accepted that, observing from above the north pole, the Earth moves anticlockwise round the Sun. The orbital plane of the Earth is called the plane of the ecliptic and the Earth's equatorial plane is inclined to this at an angle of 23° 27′ (Figure 2.5). The line of intersection of these two planes defines an important direction in space, known as the first point of Aries. At any particular time this direction is exactly known and all celestial directions can conveniently be referred to it. The advents of spring and autumn, when the Earth is at the line of intersection of the planes (Figure 2.5), are known as the vernal and the autumnal equinox, respectively. The position of the orbital plane of an artificial Earth satellite can be defined relative to the line of intersection of the Earth's equatorial plane and the ecliptic plane.

Figure 2.5. The Earth's orbit showing the directions of its revolution round the Sun and its rotation about its axis

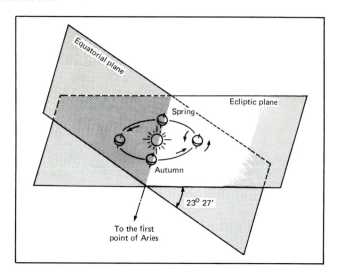

Figure 2.6. Some satellite orbital parameters

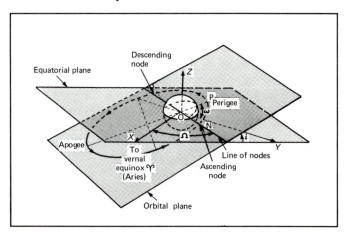

The position of the orbital plane in space is usually given in a terrestrial–sidereal rectangular co-ordinate system. The axes of such a system are shown in Figure 2.6. The origin O of the co-ordinate system is the centre of the Earth; the Z-axis is oriented towards the north pole and the Earth's equatorial plane is in the XY plane. The X-axis is oriented towards the vernal equinox or the first point of Aries. The points of intersection of the satellite orbit with the celestial equator are known as the nodes. For example, in Figure 2.6, N is the ascending node. The angle Ω between the X-axis and the line ON then defines the fourth orbital element and is called the right ascension of the ascending node. The longitude of a satellite's ground track at a particular time is given by the value of Ω.

The fifth and perhaps most important of the orbital elements is the orbital inclination, an angle i between the orbital plane (Figure 2.6) of the satellite and the equatorial plane of the Earth. Together with Ω, i specifies the position of the orbital plane relative to the coordinate system considered. The orientation of the ellipse within the orbital plane is given by the sixth orbital element, the angle between the line ON in the equatorial plane and the line OP, where P is the perigee. The line OP lies along the major axis of the ellipse. This angle, usually denoted by ω, is called the argument of the perigee.

Of these elements, i is the only one which remains practically constant during the lifetime of the satellite; others change and these changes are termed perturbations. The sources and the types of these orbital perturbations are described in the sections below.

II. Orbital perturbations

Effects of the Earth's rotation

In the case of the rotation of the Earth, since there are no real forces involved, the orbital perturbation is an apparent one. The Earth rotates on its axis at the rate of about 15° per hour. For a terrestrial observer, therefore, the satellite orbit appears to shift continually westward by about 15° per hour. This effect is discussed further in Chapter 3.

Relativistic effects

For an elliptical orbit the general theory of relativity predicts that there will be a small, continuous rotation of the line of apsides (Figure 2.4). This results in the movement of the perigee. However, the effect is very small, of the order of a few seconds per year, even for a highly eccentric orbit [1]. The relativistic effects are, therefore, usually neglected.

Effects due to the asymmetry of the Earth

Because of the rotation of the Earth round its axis, it bulges out at the equator and its poles are flattened. The shape of the Earth is rather more like a pear than a sphere. This results in anomalies in the gravitational field, for example, in the western Pacific, the Indian Ocean and Antarctica. Therefore, the orbit of an artificial satellite is distorted from that calculated when it is assumed that the Earth is spherical and that it can be represented by a point mass.

Figure 2.7. Mathematical treatment as harmonics of anomalies in the Earth's gravitational field, resulting from the slightly pear-shape of the Earth. Here the second, third, fourth and fifth harmonic deformities are shown – although not drawn to scale.

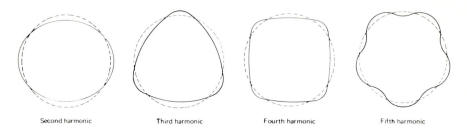

Second harmonic   Third harmonic   Fourth harmonic   Fifth harmonic

Mathematically, the problem is treated by expanding the gravitational potential function in a series of harmonics. Various terms in the expansion can be identified with certain deformations in the Earth's shape (Figure 2.7). The Earth's equatorial bulge has two effects. In one, the orbital plane of the satellite is rotated slightly farther westward on each revolution than would be expected from the rotation of the Earth alone. This changes Ω.

In the second effect, the satellite is deflected as it approaches a perigee point near the equator. The satellite slightly overshoots the previous perigee point as it swings down towards the perigee and at the same time it is pulled towords the equator. The net effect is to rotate the line of apsides and, therefore, ω.

Gravitational effects of the Sun and Moon

For satellites close to the Earth, the gravitational fields of the Sun and Moon can be disregarded in comparison with the Earth's gravitational field. However, with a highly eccentric elliptical orbit, the forces of gravity of the Sun and the Moon become significant. One effect of these forces may be to depress the perigee, thereby shortening the lifetime of the satellite. Orbital perturbations involving the satellite, the Earth and the Moon or the Sun are difficult to generalize. Each case has, therefore, to be dealt with separately.

Effects due to the Earth's atmosphere

This effect, known as atmospheric drag, causes the satellite to decelerate so that it loses some of its kinetic energy which is converted into heat. As the satellite slows down, its centrifugal force decreases and the Earth's gravity pulls it farther into the atmosphere.

The atmospheric drag force is variable. For example, the Sun significantly expands the atmosphere on the sunlit side of the Earth because of its heat and therefore causes different atmospheric drag force from that experienced in the atmosphere on the dark side of the Earth. Moreover, the Sun injects large quantities of matter into the upper atmosphere. This follows an 11-year cycle pattern. The drag force is made more complex by the fact that the Sun also injects particles during solar storms. In addition to these effects, there is a semi-annual plasma effect that causes drag to peak around April and June each year [2].

Effects of magnetic drag

When a large satellite moves through the Earth's magnetic field, an electro-motive force (e.m.f.) is established at right angles to the field and to the direction of the satellite's motion [3]. This electromotive force tends to concentrate the electrons to one end of the satellite and, in effect, polarizes it. The electrons in the upper atmosphere are attracted to the positively charged end of the satellite, whereas the heavier particles and ions are not significantly affected. This asymmetry of charge flow creates a current flow across the satellite body which is moving in the Earth's magnetic field. The result of the interaction between the current and the magnetic field is to slow down the satellite. The magnetic drag is proportional to the cube of the dimensions of the satellite [3]. At high altitudes (> 1200 km) the magnetic drag can exceed aerodynamic drag.

Effects of solar-radiation pressure

Radiation from the Sun reaching the surface of a satellite may either be reflected or absorbed by the surface. A small amount of energy is transferred to the satellite, producing a certain amount of pressure. If the radiation strikes the surface normally and if all of it is reflected, pressure created at the Earth's orbit is 9.2×10^{-6} N/m^2 [4]. Such pressures could cause considerable changes (hundreds of kilometres) in the perigee of a large but low-density satellite. Therefore, the solar-radiation pressures are only important for balloon-type satellites such as the US Echo.

Orbital perturbations caused by these factors may be undesirable. For example, atmospheric drag may prematurely terminate the lifetime of the satellite. On the other hand, such natural perturbing forces and artificial forces are frequently used to exploit the potentials of satellites. This will become apparent from the following chapters in which various military applications of satellites are described.

3. Reconnaissance satellites

Compared with aerial reconnaissance from such aircraft as the U-2, satellite reconnaissance has many advantages. For example, the high speed of a satellite – almost 30 000 km/h – enables it to survey very large areas of the Earth in a short time. Because of the relatively high altitude of, for example, a photographic reconnaissance satellite, an area of several thousands of square kilometres can be photographed on a single frame of film. In this chapter two types of reconnaissance satellites are considered – photographic and electronic.

I. Photographic reconnaissance satellites

Military reconnaissance using satellites is accomplished with the aid of sensors operating in various regions of the spectrum of electromagnetic radiation. Instruments used in photographic reconnaissance satellites are sensitive to electromagnetic radiation with a wavelength of between 0.004 mm and 0.007 mm and in the infra-red region (wavelength between 0.3 mm and 3000 mm) (Figure 3.1).

Figure 3.1. The electromagnetic spectrum. The difference between the visible light or radio waves and other radiations is that the latter have shorter wavelengths, and higher frequencies and energies.

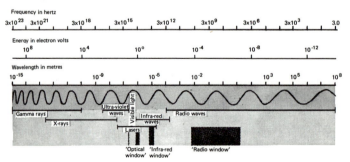

It has been mentioned above that of all the orbital elements, i, the orbital inclination, is the only element which remains practically constant. The choice of a particular value of i and other elements determines the region of the Earth over which a satellite flies. The coverage of the Earth by a reconnaissance satellite is, therefore, best studied by calculating its ground tracks, that is, the path traced on the surface of the Earth, over which the satellite flies.

Satellite ground tracks

Satellite ground tracks can be calculated using the following equations [5]:

$$L = \sin^{-1}(\sin i \sin \phi) \tag{3.1}$$

$$\Lambda = \tan^{-1}(\cos i \tan \phi) + \Lambda_{no} - \Omega_e \Delta t$$
$$- \tfrac{3}{2}KnJ_2\Delta t(R_e/a)^2 \cos i/(1 - e^2)^2 \tag{3.2}$$

where

L = geocentric latitude,

Λ = geocentric longitude,

i = orbital inclination (measured positive from east at ascending node),

ϕ = orbit central angle between the satellite and the ascending node,

Λ_{no} = initial geocentric longitude of the ascending node $(0 \leq \Lambda_{no} \leq 360°$, measured positive towards the east),

Ω_e = rotation rate of the Earth, $4.178\,074 \times 10^{-3}$ deg/s,

Δt = time measured from the initial condition,

K = 57.2958 deg/rad,

n = mean motion = $\sqrt{(\mu/a^3)}$,

μ = gravitational constant of the Earth, $398\,601.5$ km^3/s^2,

J_2 = 1.0823×10^{-3},

R_e = equatorial radius of the Earth,

a = orbit semi-major axis, and

e = orbit eccentricity.

Equations (3.1) and (3.2) neglect second-order oblateness perturbations. The equations could be used to generate the ground tracks as a series of points L, Λ as a function of the parameter ϕ. This determination is simple in the case of circular orbits but rather complicated for elliptical orbits.

13

Satellite orbits

Although a satellite's motion round the Earth is complicated by the rotation and the shape of the planet, these complicating phenomena have been effectively used to make observations from space. For example, the Earth rotates round its axis approximately once every 1440 min (that is, every 24 h). If the period of a satellite is chosen carefully (as a multiple of 24 h), then the ground tracks will repeat each day. On the other hand, a period can be chosen to result in a gradual shift of the ground tracks each day so that complete coverage of a large area of interest can be made. The ground tracks will repeat over a particular area at every sixteenth orbit if the period is close to 90 min.

A suitable choice of satellite period makes it possible to observe an area on the Earth at least once a day during daylight, or more often depending on the latitude of the target on Earth. From Figure 2.6 (see page 8), it can be seen that the orbital inclination determines the range of latitudes over which the satellite travels on each revolution. For observations of an area situated at high latitudes, a near-90° inclination orbit produces two or more daylight passes per day over the area, whereas coverage of an area near the equator requires a low-inclination orbit. Therefore, the choice of a particular i depends on where the area of interest is and how closely it is to be observed.

A second factor that might influence the choice of i is economics. If a satellite is launched eastwards from any place along the equator, it already has an initial velocity of 1700 km/h because of the Earth's rotation. Therefore, less energy is needed from a rocket to put a satellite in orbit with an i of considerably smaller value than 90°. A further advantage of a small i is that for a given latitude (except at the equator) the ground tracks are closer together than those obtained with the orbital inclination of 90°.

With such knowledge of satellite orbits and ground tracks, it is possible to learn which regions of the Earth can be observed by a satellite. Of particular interest in this connection are satellites orbited so that their ground tracks cover conflict areas at crucial times. It can only be noted that this use of reconnaissance satellites is not the accepted use stipulated in the SALT I agreements, and that military use could potentially be made of the information gained from such satellites.

Consider two satellites launched in 1974, one by the United States at an orbital inclination of 94.5° (Big Bird satellite 1974-20A) and the other launched by the Soviet Union at an orbital inclination of 62.81° (Cosmos 653) (see the initial orbital characteristics of these and other US and Soviet photographic reconnaissance satellites in Tables 3A.1 and 3A.2, respectively). The ground tracks over India of the two satellites are shown in Figures 3.2 and 3.3. It can be seen that, on consecutive days, the distances between the ground tracks of the US satellite with an orbital inclination of 94.52° are greater than those for the Soviet satellite which was launched in an orbit with the smaller orbital inclination of 62.81°. Although the former satellite, because of its high value for i, can observe the areas of the Earth between latitudes 85.5°N and 85.5°S, the latter satellite has the advantage of being able to observe an area between

Figure 3.2. Ground tracks over India of the US 1974-20A Big Bird satellite, launched on 10 April 1974 at an orbital inclination of 94.52°, 16–25 May. The date and orbit number are indicated for each ground track. Note the site of the nuclear test (starred).

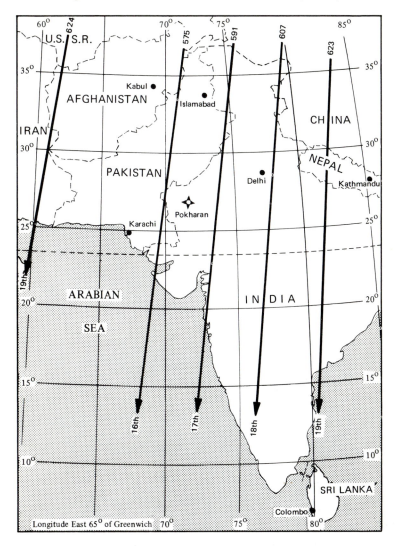

the latitudes 62.8°N and 62.8°S in greater detail since the corresponding ground tracks on consecutive days are much closer together. It can be seen from the two ground tracks made on 19 May for the US satellite (Figure 3.2) and by the two ground tracks on 25 May for the Soviet satellite (Figure 3.3) that they are about $22\frac{1}{2}°$ apart, a result of the orbital period of nearly 90 min.

So far, the gradual shift of the ground tracks each day due to the satellite's period not being an exact sub-multiple of the Earth's period of rotation round its axis has been considered. There is a second effect which causes shifts in the

Figure 3.3. Ground tracks over India of the Soviet Cosmos 653 satellite, launched on 15 May 1974 at an orbital inclination of 62.81°, 16–25 May. The date and orbit number are indicated for each ground track. Note the site of the nuclear test.

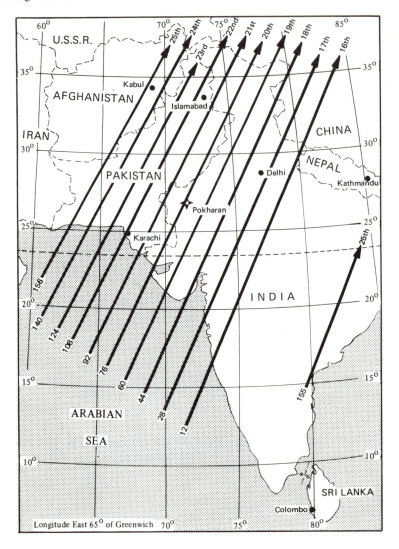

satellite ground tracks due to the changes in other orbital elements. Basically, three orbital elements change their values, causing further shifts in the ground tracks. First, the orbital period of a satellite varies because of, for example, the Earth's atmosphere. The changes in the other two orbital elements Ω and ω are caused by the Earth's uneven gravitational field. This deviation of the Earth's shape from a perfect sphere causes the plane of the satellite orbit to rotate round the Earth's axis, so that the value of Ω changes with time, while keeping the orbital inclination constant. This change can amount to as much

16

as 10° per day. The effect is known as precession. The Earth's uneven gravitational field causes the ellipse to rotate in its own plane so that ω changes with time. This change can amount to as much as 5° per day. However, at an orbital inclination of exactly 63.4° the perigee remains stationary. From Figure 2.6, page 8, it can be seen that ω determines the latitudes in which the perigee and apogee are situated. Apart from the effects of these changing orbital elements, the Earth's period of rotation – slightly less than 24 h – also causes a small shift in the ground tracks.

For reconnaissance satellites the change in Ω is used so that they pass over a region of interest on the Earth at the same time of day throughout the active lifetime of the satellite. Thus the reconnaissance photography always refers to the same local time so that changes in activity in the region can be compared from day to day. This is done by placing a satellite in a Sun-synchronous orbit in which the satellite is orbited with an inclination of almost 90°. The plane of the satellite orbit contains both the Earth and the Sun. In this type of orbit the satellite crosses the equator at just about local noon on the sunlit side of the Earth and local midnight on the dark side. The Earth rotates under the satellite orbit, which is fixed in space, at a rate of 15° per hour, so that the equatorial and temperate zones of the Earth can be photographed with the Sun always high in the sky. However, the Earth rotates around the Sun and since the satellite orbital angle i is fixed, a quarter of a year later the satellite orbital plane will be perpendicular to the plane containing the Earth and the Sun. To maintain the Sun-synchronous orbit, the plane of the satellite orbit has to be rotated 360/365 or 0.986 degree per day. If the satellite is in just the right orbit, the Earth's equatorial bulge will deflect Ω and therefore the satellite orbit by this amount. The US 1974-20A satellite over India described above was orbited in just this type of Sun-synchronous orbit.

The use of the Sun-synchronous orbit is a deliberate use of the changing character of Ω. However, the combined effect of the Earth's rotation and the changing orbital elements causing shifts in ground tracks can also be used to advantage. Examples of this type of use include the Soviet Cosmos 666 satellite launched on 12 July 1974 at an orbital inclination of 62.81° and a close-look type of US satellite (1974-65A) launched on 14 August 1974 at an orbital inclination of 110.51°. The ground tracks of these satellites are shown in Figures 3.4 and 3.5, respectively. The ground tracks made by the Cosmos satellite on 14 July and 24 July could have coincided but because they shifted, the latter track just managed to come over eastern Cyprus. There are several other such closely located pairs of ground tracks covering various parts of Turkey. Again, it can be seen in Figure 3.5 that a number of such pairs of ground tracks have been made by the US satellite over Greece and Turkey. Another interesting point about this satellite is that passes were made twice a day over most of the area shown in the figure.

The dates when these satellites orbited over Cyprus, Greece and Turkey are interesting, particularly for Cosmos 666. On 15 July 1974 there was an army coup in Cyprus and between 20 and 22 July Turkey invaded Cyprus. The Cosmos satellite was near western Cyprus on the morning of 15 July (around

Figure 3.4. Ground tracks over Cyprus, Greece and Turkey of the Soviet Cosmos 666 satellite, launched on 12 July 1974 at an orbital inclination of 62.81°, 13–25 July. Passages were made during the morning. The date and orbit number are indicated for each ground track.

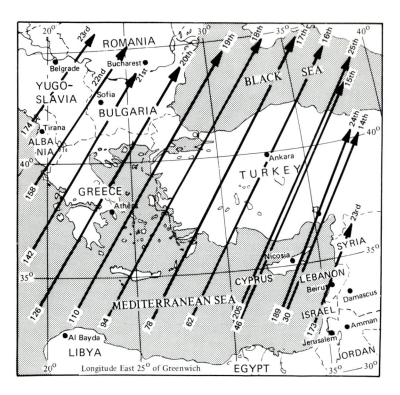

1000 hours local time) and over the eastern region on the afternoon of 22 July (about 1620 hours local time). The following day, the satellite passed over western Cyprus (at about 1600 hours local time), and on 24 July it passed over eastern Cyprus (at about 0700 hours local time). Although telemetry [6] from Cosmos 666 suggested that it had a manoeuvring capability, no manoeuvre was detectable. The US 1974-65A satellite made only two passes over Cyprus on 22 August, which was after the Turkish invasion but not very long after the occupation of northern and north-eastern Cyprus. Reports on the cloud coverage based on observations made both from meteorological satellites and from ground stations indicated that throughout these periods in July and August, the sky over Cyprus was free of clouds. During these periods the sky over Greece and Turkey was also mostly clear.

These changes in ground tracks due to natural causes are slow, however, and it often takes a long time before a satellite is correctly positioned over an area of interest. Satellites are therefore often manoeuvred from a ground station on the Earth in order to bring them over a specific region more quickly. Such manoeuvring is best illustrated by a Soviet satellite, Cosmos 667, which was launched on 25 July 1974 (see the ground tracks in Figure 3.6). It can be

Figure 3.5. Ground tracks over Cyprus, Greece and Turkey of the US 1974-65A satellite, launched on 14 August 1974 at an orbital inclination of 110.51°, 15–30 August. Southbound passages were made during the morning and northbound passages during late afternoon. The date and orbit number are indicated for each ground track.

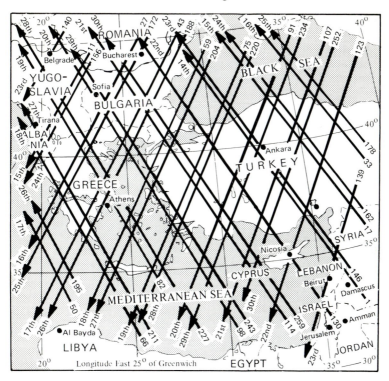

seen that, until 2 August, the ground tracks were relatively widely spaced covering Cyprus, Greece and Turkey. But after 2 August the satellite was manoeuvred in order to change some of its orbital elements so that its ground tracks became very narrowly spaced and concentrated over Cyprus. The satellite made two daytime passes (about 0730 hours and 1630 hours local time) per day for four days between 3 and 7 August (Figure 3.6), and it was probably re-covered on 7 August. Again, weather reports from meteorological satellites and ground stations indicated that sky over Cyprus during these five days was very clear.

It is interesting to note that on 15 July, when the army coup occurred in Cyprus, the US satellite 1974-20A (Big Bird) flew over Cyprus (Figure 3.7) at about 1025 hours local time. The satellite made two more passes over Cyprus: on 20 July when the Turkish invasion began and on 24 July. On both these occasions the satellite orbited over Cyprus at just after 1000 hours local time, when the Sun was at about 60° in the nearly cloud-free sky. These passes over Cyprus were probably made by manoeuvring the satellite, but this is not easy to determine from the orbital elements since the prolonged life of Big Bird satellites is normally obtained by changing these elements artificially.

19

Figure 3.6. Ground tracks over Cyprus, Greece and Turkey of the Soviet Cosmos 667 satellite, launched on 25 July 1974 at an orbital inclination of 64.98°, 25 July–7 August. The ground tracks show that the satellite was manoeuvred after 2 August. The date and orbit number are indicated for each ground track.

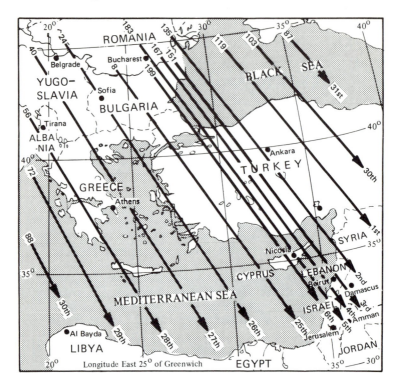

In general the selection of a particular type of orbit for a photographic reconnaissance satellite depends on some of the following factors. For most missions, orbits with as small an eccentricity as possible are selected so that uniform conditions are obtained for the sensor on board the satellite. For low-altitude satellites, the orbital inclination must be at least as large as the latitude of the geographical area to be observed. The choice of a particular satellite altitude must be a compromise between high sensor resolution and low power requirements for low-altitude satellites and larger coverage and satellite orbital lifetime obtained for high-altitude satellites. In order to achieve complete coverage of an area of interest, an orbital period close to 90 min must be chosen.

Space photography

A photographic reconnaissance satellite orbiting at an altitude of 150 km can view an area nearly 18 times larger than that seen by an aircraft flying at an altitude of about 9 km; aircraft at this altitude can view about 290 000 km². A

Figure 3.7. Ground tracks over Cyprus, Greece and Turkey of the US 1974-20A satellite, launched on 10 April 1974 at an orbital inclination of 94.52°, 14–28 July. The date and orbit number are indicated for each ground track.

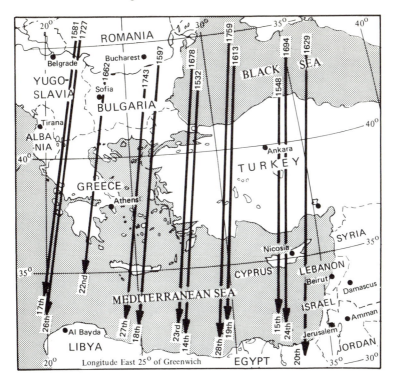

further advantage in using satellites is that humans are not exposed to direct risk. During the past four years or so, it has been observed that photographic reconnaissance satellites have been routinely used to monitor conflict areas of the Earth. It is perhaps appropriate at this stage to look at the resolution of photographic images obtained from such satellites.

The overall performance of a photographic system depends on several factors: (a) the contrast between the object being photographed and its surroundings; (b) the shadow cast by the object being photographed; (c) the satellite's altitude, which varies from perigee to apogee; (d) the atmosphere, which modifies the light reaching the satellite camera system; (e) the characteristics of the camera carried by the satellite; and (f) the type of film used to record the image. The details of what can be seen by reconnaissance satellites are still a very closely guarded secret, but a theoretical treatment of the capabilities of the photographic equipment on board the satellite involves the above six factors.

However, some estimates of the image quality of photographs taken from reconnaissance satellites can be made from a study of recently published photographs taken from Landsat 1 and 2 (Earth Resources Technology Satellites) and the Skylab space station. In order to estimate the quality of an image, an

21

Figure 3.8. Simple geometry of the satellite camera system

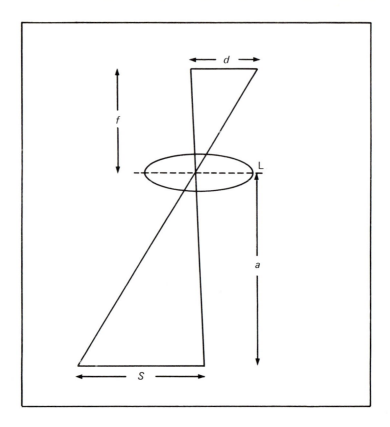

important basic parameter is resolution. The concept of resolution is derived from the optical criterion which defines it as the minimum distance between two point-sources of light which can still be distinguished as separate points. Photographic resolution is defined as the minimum observable spacing between black and white lines in a standard pattern. Therefore, a photographic resolution of 10 lines per millimetre means that black and white lines, both 0.1 mm wide, are just barely resolvable. A more common term of resolution used in connection with reconnaissance photographs is the ground resolution, which is defined as the ground dimension equivalent to one line at the limit of resolution.

The theoretical treatment of ground resolution is complicated since it involves not only the properties of the lens system of a camera but, also, among other things, the characteristics of the film used. However, an approximate expression for the ground resolution can be derived as follows. Consider Figure 3.8 in which L represents the camera lens system with a focal length f metres placed on a satellite at an altitude of a kilometres. The object on the ground is represented by S metres so that its image in the image plane of L is d metres. If d is the smallest size image that can be resolved, then, by definition, S is the ground resolution. By simple geometry, the ground resolution is then given by

Figure 3.9. Ground resolution as a function of focal length (satellite altitude = 150 km)

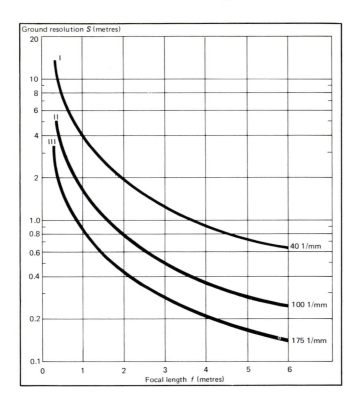

$$S = (a/f)d. \tag{3.3}$$

It is common to use the reciprocal of d and call it the limiting resolution R of the system. R is the combined resolution in lines per millimetre of the camera lens system and the film. The ground resolution is then given by

$$S = a/fR. \tag{3.4}$$

For various values of R, ground resolutions for a number of focal lengths have been calculated and plotted in Figure 3.9. The satellite altitude of 150 km was used for these calculations since this is the average perigee height of US reconnaissance satellites.

The resolution obtained in a photograph of the Chicago area from the Skylab space station is shown in Figure 3.10. The photograph was taken with a 460-mm focal-length lens. Although it was taken from an altitude of about 440 km, it must be noted that most US reconnaissance satellites orbit at an altitude of about 150 km. At this altitude, the ground resolution for the Skylab system would be $25 \times 150/440$ or about 8.5 m, as opposed to the resolution of 25 m illustrated in the plate. The photograph clearly shows such landmarks as highway intersections and airport runways.

Figure 3.10. The Chicago area photographed from the Skylab space station, from an altitude of about 440 km:
(1) Highway intersections Illinois Routes 51 and 180, near LaSalle and Perie, and Routes 180 and 155, west of Jobet; (2) Lakeport, Mergs Field, and Midway Airport; (3) Main ring of the accelerator at the Fermi National Accelerator Laboratory (its outer diameter is 2 km); (4) O'Hare International Airport and runways (A) (the runways are about 45 m wide and 100 m apart).

From Figure 3.9 it can be seen (curve I) that the ground resolution of 8.5 m could be obtained from a camera with a focal length of 460 mm if the total resolution (that is, the lens plus the film) is 40 lines per millimetre. But modern film materials have considerably higher resolutions than 40 lines per millimetre, as is illustrated in Figure 3.11, which shows the US MacDill Air Force Base. Four aircraft standing off the main runway can be seen encircled in an enlargement of a small section (about 5 mm square) of a photograph taken from NASA's Skylab satellite; the complete frame (about 9 cm square) shows Tampa Bay, Florida. With cameras equipped with lenses of long focal length, it is not unreasonable to envisage an image resolution of at least 100 lines per millimetre. With a resolution of this order and with a focal length of 600 cm, it is possible to get a ground resolution of 25 cm (curve II, Figure 3.9). It has even been suggested

24

Figure 3.11. An enlarged section of a photograph taken from the Skylab satellite showing the MacDill Air Force Base, Florida. Four aircraft standing off the main runway are clearly visible within the circled area.

that an image definition of 175 lines per millimetre is possible [7], in which case a ground resolution of about 15 cm can be obtained with a camera of focal length 600 cm (curve III).

It would seem from Figure 3.9 that the ground resolution could be improved considerably if the focal length of the camera were to be increased. However, the diameter of the optical system increases with increased focal length. Furthermore, the maximum size of the diameter is limited by the diffraction effect within the optical system. But even taking this into account, the diameter required to

Figure 3.12. A US Discoverer military photographic reconnaissance satellite

achieve a focal length of 600 cm is not outside the bounds of practicability. Atmospheric effects also cause a deterioration in the resolution but with advanced technology it is possible to make up for such losses in resolution by using computers.

With the possibility of such high ground resolutions as 15 cm it is not surprising to find the use of reconnaissance satellites extended to activities other than merely providing assurances of compliance to arms control treaties. Similarly, the activities of civilian satellites increasingly overlap into more military areas. Recently, for example, there were reports, based on the observations of Landsat satellites, of violations of the interim SALT I agreement and the ABM (anti-ballistic missile) Treaty [8–10].

In the case of the Soviet Union, such reports are based on the photographs taken by Landsat 1. It would be interesting to examine the kind of details that could be observed from the Landsat 1 satellite. The ground resolution of a Skylab camera is about 25 m from an altitude of 440 km, or about 50 m (25 × 910/440) at an altitude of 910 km, the normal altitude of the Landsat satellites. The ground resolution in the green and red part of the visible spectrum for a Landsat RBV (return-beam-vidicon) television camera is about 150 m [11]. This is less than the ground resolution of the Skylab camera by a factor of three, which means that the two runways of the O'Hara International Airport shown in Figure 3.10 could not be resolved by a Landsat camera but would appear as a single runway. However, in a Landsat photograph, major roads, airports and to some extent their runways could be identified, and image quality could be considerably improved with the aid of computer image-enhancement techniques.

26

Figure 3.13. The nose cone of a Discoverer satellite similar to one which was used in recovering a capsule from orbit round the Earth

Further developments have made photography possible under very low light intensity, such as moonlight or even starlight [12]. Illumination of target areas by laser beams has also been developed. Developments in television cameras have meant that pictures can be transmitted back to Earth in a relatively short time. New types of vidicon tubes for such cameras give resolutions which approach those of ordinary photographic cameras [13, 14]. The advances in synchronous communications satellites which can handle television transmission bandwidths would not only make it possible to relay reconnaissance photographs to ground stations but would also enable ground operators to switch to a lens of different focal length on the satellite camera or to insert filters of different colours so as to detect objects under camouflage. In addition to optical and television cameras, developments in sensors sensitive to radiation outside the range of the visible spectrum broaden the range of uses of satellites. Further, side-looking radar, giving a useful resolution, allows reconnaissance to be carried out under all weather conditions. A conventional radar can easily penetrate clouds but even with a large, three-metre diameter antenna, an object of 1.5 km in size could not be easily resolved from an altitude of about 200 km. Side-looking radar can give better resolution using a smaller antenna, provided that the antenna is aimed to the side rather than forward, as is the case with a conventional radar [12, 15].

The US programme

The use of artificial Earth satellites for reconnaissance purposes was recognized in the United States as long ago as 1946. By 1954, considerable details were

27

Figure 3.14. Vandenberg Air Force Base (WTR), showing launch complex areas

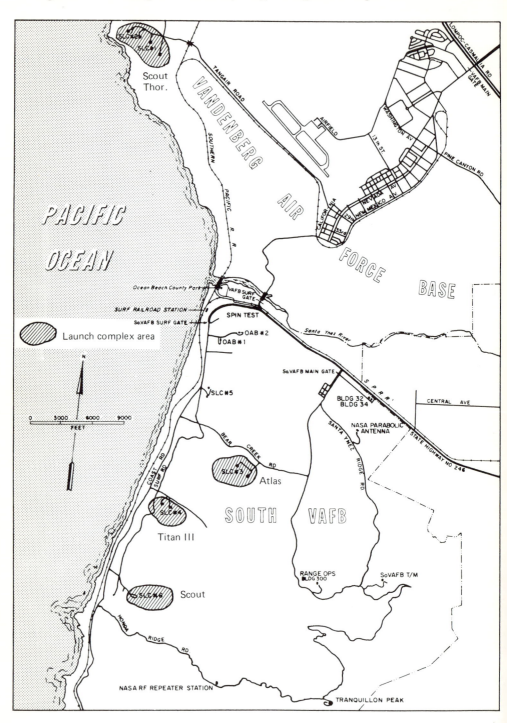

worked out on these satellite systems. In 1955, detailed descriptions of technical requirements were submitted by the United States Air Force (USAF) to various industrial firms, with requests for design proposals. The programme was designated Weapon System 117L (WS-117L) and in October 1956 the development of the Agena rocket, which has since been the basic launch vehicle for reconnaissance satellites, began.

In November 1958 the US Department of Defense (DoD) officially disclosed that WS-117L consisted mainly of three separate programmes – Discoverer, which was concerned with research and development of, among other things, photographic reconnaissance techniques; the Satellite Missile Observation System (SAMOS), a photographic reconnaissance satellite system; and the Missile Defense Alarm System (MIDAS). The SAMOS designation was discontinued by the USAF in 1961 and subsequently US photographic reconnaissance satellites were not officially identified. A Discoverer satellite is shown in Figure 3.12 and its nose cone in Figure 3.13.

Most of these satellites have been launched from Vandenberg Air Force Base at Point Arguello (WTR), about 240 km northwest of Los Angeles on the west coast of the USA (see Figure 3.14). These satellites, with perigee heights of between 120 and 370 km, have two basic missions, each involving different orbital lives for the satellites. The first type of mission requires an area-surveillance satellite, its purpose being to scan a large area of a particular country for objects of potential interest. The satellite therefore carries a wide-angle, low-resolution camera. This type of satellite usually has an orbital life of three to four weeks. When the satellite is within communication range of one of the Air Force ground stations, the exposed film, already developed aboard the spacecraft, is scanned by electronic devices and the resulting electrical signals are transmitted to Earth by radio. At the end of the mission, the satellite re-enters the Earth's atmosphere and burns up.

The US reconnaissance satellite programme actually began in 1959 with the launching of six Agena-A satellites using Thor launchers which are capable of putting payloads of 600–900 kg into orbit. The payload was increased during subsequent years, with the use of Thor/Agena-B and Thor/Agena-D combinations, which are able to carry greater payloads of 1000–1500 kg. By 1964–65 improved versions of booster rockets, including the so-called Thrust Augmented Thor (TAT) and the Long Tank TAT (LTTAT), were developed and can carry payloads of between 1600 and 2000 kg [13, 15]. The development of these various satellites and launchers is shown in Table 3.1.

The second type of reconnaissance mission uses the 'close-look' satellite, which carries a camera with high resolution and relatively narrow field of view. The purpose of this mission is to re-photograph areas of particular interest located by the area-surveillance satellites. Close-look satellites are larger than those used for the first type of mission and they remain in a near-polar orbit for about five days before the film itself is recovered. Recently developed satellites remain in orbit for longer periods. At the end of the mission a capsule (Figure 3.15) containing the film is ejected from the spacecraft and in some cases recovered when it lands at sea within a predetermined area. More often, the

Table 3.1. US photographic reconnaissance satellites, launchers and orbital inclinations

| Year | Area-surveillance satellites | | | | | Close-look satellites | | | | Big Bird satellites |
	Th/A-A	Th/A-B	Th/A-D	TAT/A-D	LTTAT/A-D	Atlas/A-A	Atlas/A-B	Atlas/A-D	Titan-3B/A-D	Titan-3D
1959	6									
1960	4	2								
1961		11				1				
1962		14	6				1			
1963			8	7			6			
1964			4	11				4		
1965				13				9		
1966				8				8	3	
1967				3	6			12	6	
1968					8			3	8	
1969					6				6	
1970					4				5	
1971					2				4	1
1972					2				3	3
1973									3	2
1974									3	2
1975									2	2
1976									2	2
Orbital inclination	65°, 70°–75° and 80°–85°					90°–110°			105°–115°	95°–100°

Figure 3.15. A Discoverer satellite capsule

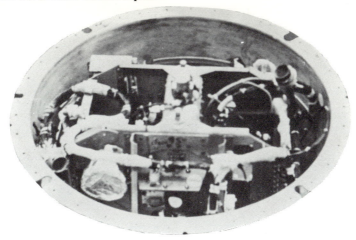

Figure 3.16. Re-entry of capsule from a US reconnaissance satellite

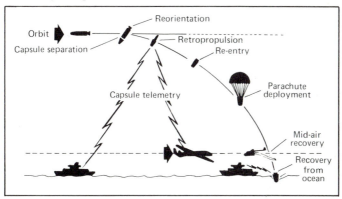

capsule is recovered in mid-air (Figure 3.16); when the capsule reaches an altitude of about 15 km, a parachute fixed to it opens and is caught by a trapeze-like cable attached to either a C-130 transport aircraft or to a recovery helicopter. The aircraft is guided by ground radar and radio-beacon signals from the capsule. The films, thus recovered, are then developed and analysed.

Initially, Agena photographic reconnaissance satellites with close-look types of mission were launched using Atlas launchers (Table 3.1); from mid-1966, Agena-D satellites have been launched using Titan-3B rockets. Although both these launchers are the same size, the payloads put into orbit differ: 1500–2000 kg for Atlas and about 4500 kg for Titan-3B launchers [13]. The increased payload capability allows larger film packs, longer focal-length cameras and a larger number of film-recovery capsules to be used.

A satellite system in which one type of satellite identifies potentially interesting areas and another makes detailed studies of them entails certain disadvantages since a large number of photographs must be taken by the surveillance satellite, many of which may be of no interest. Moreover, the recovery

Figure 3.17. A Titan-3D rocket launching a Big Bird reconnaissance satellite into orbit

and processing of these photographs introduce some delay before a close-look satellite can be launched. A close-look satellite is usually launched about four to eight weeks after the launch of an area-surveillance satellite. This has led to the development under Program 467 (or LASP – Low Altitude Surveillance Platform) of the third type, the new-generation Big Bird satellite (Figure 3.17), designed to perform both the area-surveillance and the close-look types of mission. Big Bird satellites are launched using Titan-3D rockets which can place a payload of some 13 600 kg into low Earth orbit. The vehicle consists basically of two Titan core stages with two five-segment strap-on solid boosters three metres in diameter. The satellite has its own propulsion system for orbital adjustments during flight.

The use of area-surveillance and close-look satellites has declined and in fact, since 25 May 1972, no US area-surveillance satellite has been launched. Instead, Big Bird satellites are equipped with high-resolution cameras, carry six recoverable film capsules and their photographs are usually returned to the Earth at two- or three-week intervals [6]. The spacecraft is also believed to be equipped with large antennas so that photographs taken during an area-surveillance mission can be transmitted to Earth. In addition, it is believed that Big Bird satellites carry ultra-high frequency equipment to provide com-

Figure 3.18. A full-scale model of Cosmos 381, displayed at the 1971 Paris Air Show. The satellite model was a cylinder measuring 1.4 m in height and 2 m in diameter.

munication with US Strategic Air Command aircraft operating in the polar region [6, 7].

From Table 3.1 it can be seen that the United States launched only four photographic reconnaissance satellites during 1976. Of these, two were the large Big Bird satellites and two were close-look. Fewer US photographic reconnaissance satellites have been launched in recent years because the lifetimes of these satellites are increasing almost every year. The first Big Bird satellite launched in 1971, for example, had a lifetime of only 52 days compared to the lifetime of 158 days of the satellite launched on 8 July 1976. A Big Bird satellite launched on 19 December 1976 was still in orbit after 301 days. Some eight-fold increase in the lifetime of the close-look satellite has occurred; a satellite launched on 22 March 1976 had a lifetime of 57 days compared with seven days for one launched in July 1966. This has resulted in fewer US close-look satellites being launched every year; only two were orbited in 1976 and three each in 1973 and 1974.

The characteristics of the US photographic reconnaissance satellites are given in Table 3A.1.

The Soviet programme

Interest in space flight in the Soviet Union dates back at least to the past century when, among others, K. I. Konstantinov (1817–71) began publishing his ideas on the subject and laid the scientific foundation for Soviet military missiles. The founder of space engineering, however, was K. E. Tsiolkovsky, who in 1883 described an interplanetary craft with a rocket engine and in a series

Figure 3.19. Photograph of the Tyuratam Cosmodrome in the Soviet Union, taken from a US Landsat satellite:
(1) Central complex with major administration, booster-assembly and laboratory facilities.
(2) Location of a previous and smaller space/military booster complex.
(3) Location of a new space launch complex (under construction when photograph was taken).
(4) A new R&D test complex.
(5) An above-ground fuel storage area.
(6) The town of Tyuratam.
(7) Railway lines which support the centre and transport the booster to the launch sites.
(8) Syr Dar Ya River.

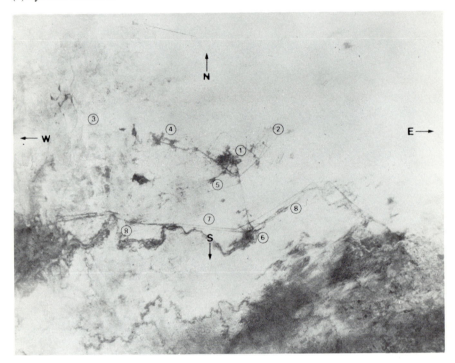

of publications between 1903 and 1914 formulated the theory of rocket flight and the basic principles of space rocket designs [18]. But it was during the mid-1960s that the Soviet programme began to proliferate in many directions.

Most satellites launched by the Soviet Union are designated Cosmos (Figure 3.18). The only Cosmos satellites which have been identified with any degree of confidence as having a reconnaissance function are those which are recovered from orbit after periods of up to 14 days and which probably perform photographic reconnaissance missions.

A number of other characteristics of the Soviet satellites, such as the narrow range of orbital inclinations used, make their identification as reconnaissance

Figure 3.20. Drawing made from a photograph taken from a Landsat satellite of the Soviet northern launch site.

Key: (1) the town of Plesetsk which is both a support centre as well as a housing facility for the personnel – the launch centre is about 100 km long; (2) launch complexes; (3) the main airport with a single runway and a single taxiway; (4) and (5) are two additional launch complexes; (6) part of a rail and road complex; (7) a ground test area; (8) a secondary support base; (9) tracking and communication centres; (10) two support centres, and (11) a surface-to-air missile site.

satellites relatively simple. Moreover, a certain amount of knowledge has been acquired about the nature of the radio signals transmitted by these satellites, which facilitates their identification [19, 20].

Soviet reconnaissance satellites are launched either from Tyuratam (45.6°N, 63.4°E) about 160 km east of the Aral Sea (Figure 3.19) or from Plesetsk (62.9°N, 40.1°E) about 1000 km north of Moscow (Figure 3.20). The accurate location of Tyuratam was publically known in 1957 following the observations made of Sputniks 1 and 2 [21]. Very few non-Soviets have, in fact, seen this site; however, in 1970, President Pompidou viewed the launching of Cosmos 368, a photographic reconnaissance satellite. The second site near the town of Plesetsk has never been specifically acknowledged. This site has been increasingly used and is now the busiest of all launch sites in the world. It is used primarily as an operational site, in contrast to the experimental or specialized nature of the Tyuratam flights. The first widespread public knowledge of this site came from the former Kettering Grammar School in the UK in 1966 [22].

In both the USSR and the USA, development in long-range missiles essentially provided most of their space launch vehicles. An essential difference in the development of the space vehicles in these countries was that whereas the USSR initially launched its spacecraft with ICBMs, the USA first launched its satellites using nonmilitary launch vehicles. The main basic launch vehicle in the Soviet Union is the SS-6 Sapwood ICBM (basic type A launchers), first introduced in 1957. Initially the vehicle was capable of orbiting a 360-kg payload into low orbit. The vehicle has been upgraded with the addition of a more powerful $2\frac{1}{2}$ stage giving its present capability of orbiting 6757 kg into low orbit. Soyuz (type A-1) or Voskhod (type A-2) launch vehicles have been used to put the Soviet reconnaissance satellites into orbit. The normal payload is between 4000 and 6000 kg. Each satellite probably consists of a spherical capsule with its recovery parachute, an instrument package and a location-beacon transmitter. The instrument package is separated from the capsule during the last orbit and burns up in the Earth's atmosphere; the conventional signals cease with the destruction of the instrument package [19]. The location beacon of the capsule, which continuously transmits in Morse code one of the four pairs of letters TK, TG, TF or TL, is switched on as the capsule approaches the Earth. The strength of the signals abruptly decreases when the capsule has reached the ground and the signals stop when recovery is made [23].

Soviet photographic reconnaissance satellites launched between March 1962 (when the Cosmos series began) and December 1976 are listed in Table 3A.2. Until about the end of 1968, most of these satellites had orbital lives of about eight days. A new generation of satellites, beginning with the launching of Cosmos 208 in March 1968, had longer orbital lives of about 12 days. This change to longer-duration flight is associated with the improvement in resolution. Better resolution necessitates the use of cameras with longer focal-length lenses which results in a smaller area of the Earth being photographed on each pass. Therefore, shorter orbital periods are necessary if the satellite ground tracks are to be closely spaced, which can be achieved by reducing the apogee heights. It can be seen from Table 3A.2 that these orbital criteria apply to 12-day satellites. Both the 8- and the 12-day satellites transmit information in pulse duration modulation (PDM) form [20].

A further development – satellite manoeuvrability – marked the beginning of yet another type of satellite. First used in Cosmos 251, this facility to change the orbital characteristics increased the ability of the satellite to achieve precise coverage of specific areas. These satellites used to fly for 13 days and transmit groups of Morse characters [20]. Such satellites with manoeuvring capability are equivalent to the US close-look satellites.

A further characteristic of Cosmos 251 was that on the twelfth day of its flight it ejected two objects. Similar objects were ejected by many other Soviet reconnaissance satellites on the day preceding the recovery of the main satellite [23]. Such objects are believed to be either vernier control rockets, used to make small adjustments in the orbital period of the spacecraft so as to bring it directly over targets of interest, or discarded scientific packages [24]. Cosmos 208, launched in 1968, was the first Soviet satellite to eject a scientific capsule. Of a total of 29 satellites launched in 1968, four ejected either vernier control rockets or scientific packages. In 1972, 23 of the 29 Soviet satellites ejected such objects; and in 1973, the proportion was 27 of 35 satellites. It is possible that satellites which eject such objects, and manoeuvre, are close-look satellites.

It is interesting to note that a Soviet area-surveillance satellite was launched at the beginning (7 April 1972) and at the end (21 June 1972) of a series of close-look satellites launched primarily during May and June. This sharp increase in the number of Soviet reconnaissance satellites may be related to the SALT I agreements signed on 26 May 1972.

Among the Soviet satellites launched in 1975, two area-surveillance satellites, Cosmos 720 and Cosmos 759, were dual-purpose satellites; besides performing the usual military reconnaissance missions, the satellites also conducted tasks similar to those of the US Landsat satellite [25a].

Another interesting satellite which might belong to this series was Cosmos 758. Launched from Plesetsk, it exploded after only four days in orbit. It has been implied that this satellite might have been part of the Soviet Satellite Intercept tests or that it might have been intentionally exploded after a mission failure [16]. It is possible that the satellite was on a photographic reconnaissance mission and carried a high-resolution camera [27]. It is difficult to be certain about this satellite because it was orbited at 67°, an unusual orbital inclination for a Soviet reconnaissance satellite. Two such satellites, Cosmos 805 and Cosmos 844, were launched recently: the former transmitted on a new frequency, manoeuvred during flight and was recovered after 20 days [28]; the latter exploded after three days. Certainly Cosmos 758 marked the beginning of a new programme. These satellites may be the first of the Soviet long-lived photographic reconnaissance satellites.

The Chinese programme

The People's Republic of China was the fifth nation to launch a satellite and the third to launch a reconnaissance satellite. The first Chinese satellite was launched on 24 April 1970 and was placed in a highly elliptical orbit with a perigee of 441 km and an apogee of 2380 km. The orbital period of the satellite was 114 min. The characteristics of the orbit of China 2, launched on 3 March 1971, were similar.

The first Chinese satellite with an orbit characteristic of a reconnaissance satellite was China 3, launched on 26 July 1975. However, this satellite may not have performed a reconnaissance mission. The satellite was probably launched using a modified version of the long-range (5650 km) CSS-X3. China 3 and the two subsequent Chinese satellites launched in 1975 had low perigees (of the order of 180 km). The satellites were launched by a different vehicle which permitted the presence of a camera on board China 4 in order to monitor, for example, Soviet troop movements and military installations, particularly along the Sino–Soviet border. These satellites were launched from the Shuang-Cheng-Tzu space facility approximately 1600 km west of Peking. It is interesting to note that, so far, China's is the heaviest of any state's first satellite: China 1 had a payload of about 170 kg as compared with the payload of the first Soviet satellite, Sputnik 1 (84 kg), and the payload of the first US satellite, Explorer 1 (14 kg).

The second Chinese satellite weighed about 220 kg but the weights of subsequent satellites have not been published. It is believed, however, that China 4 and China 5 weighed between 2700 and 4500 kg [29, 30]. The secrecy about the payloads and specific functions of these satellites, together with the statements made in Hsinhua News Agency reports about the satellite programme being in support of 'preparedness against war', lead one to believe that the launching of China 3 may well have been the beginning of the Chinese military reconnaissance satellite programme.

Although it was reported that China 3 was intentionally brought down [31], this is doubtful since the satellite had already been in orbit for 50 days and its orbital characteristics appeared to be those of a naturally decaying satellite [32].

Since the launch of China 3, China 4 and China 5 have been orbited, the latter with orbital parameters similar to those of China 3. It is interesting to note that, unlike that of the other Chinese satellites, the orbital inclination of China 4 was 62° and after six days a data capsule was recovered [33]; a large object remained in orbit for a further 27 days. China 4 may thus be the first Chinese reconnaissance satellite.

It is difficult to determine with certainty which of these satellites were photographic reconnaissance satellites and which were on missions similar to those of the US Earth Resources Technology Satellites (Landsat). Identification is made particularly difficult since these satellites transmit signals only when they are directly over China. China 3 might have been the first satellite with a camera on board [34] but China 4 was probably the only one to perform a photographic reconnaissance mission using a high-resklution camera since part of its payload was recovered.

In 1976, China 6 and China 7 were launched. The orbital characteristics of China 7 were similar to those of China 4. It was reported that China 7 was recovered [35] but it is possible that this was only a part of the payload.

The orbital characteristics of the Chinese satellites are given in Table 3A.3.

The French programme

France is the third nation, after the United States and the Soviet Union, to have launched its own satellite with its own rocket. Although most of the French space programme is for peaceful purposes and although the 1975 budget for the French national space agency, Centre National d'Etudes Spatiales (CNES), has shown a continuation of the shift in emphasis from a national to a European programme, in 1973 France announced interest in developing military reconnaissance satellites. Former French Defence Minister Michel Debré reported that the Defence Ministry was studying the possibility of developing a military reconnaissance satellite [36]. He emphasized, however, that it would be a long-term project which would not be implemented until sometime between 1980 and 1985.

At an exhibition at Le Bourget in 1975, the armed forces and the CNES for the first time publicly admitted interest in military photographic and electronic reconnaissance and communications satellites. The exhibition displayed, among other things, satellite systems for investigation of the Earth's resources as well as some of the other above-mentioned systems. Some details of certain of these satellites were discussed at the exhibition. For example, it has been pointed out that the photographic reconnaissance satellite would use a recoverable capsule containing the exposed films, and cost estimates of some satellites have been given. According to a study carried out by Aérospatiale-Thomson-CSF, the cost of a French communications satellite would be approximately \$750 mn–\$1000 mn, and that of an electronic reconnaissance satellite would be about twice this amount [37].

Although some have argued that France needs a military communications satellite more urgently than a photographic reconnaissance satellite [38], the French Minister of Defence, Yvon Bourges, asked the CNES and the Direction Technique des Constructions et Armes Navales (DTEN) to submit plans for the military photographic reconnaissance satellite programme to the French Parliament in March–April 1976 [39]. The satellite will be launched mainly into polar or near-polar orbit with perigee heights of about 300 km or 1200 km and will weigh about 350 kg. The higher-altitude satellites presumably will be Earth resources satellites, and will be launched using the Ariane rocket. It is planned that after the feasibility study, the armed forces will submit tenders in 1979 and launch a satellite by about 1985. The first system is expected to become operational in 1986. The satellite may be launched from the Kourou launch site in French Guiana.

II. Electronic reconnaissance satellites

The primary function of electronic reconnaissance or ferret satellites is to pinpoint the locations of air-defence and missile-defence radars usually located along a country's borders. These satellites also determine the signal character-istics and detection range of such radars so that, with a knowledge of their location, the penetration of enemy air defences by strategic bombers can be efficiently planned. In a very large country there may be many more defence radars located deep inland. Radars operate at high frequencies so that their range is limited and, because of the Earth's curvature, it is not possible to measure the characteristics of such radars in the usual way from ships and air-craft outside the country. Moreover, the characteristics of these radars are certain to be different from those at the borders. Electronic reconnaissance satellites are very useful in investigating such inland radar systems. The sensi-tivity of the sensors on board an electronic reconnaissance satellite range between millimetre-wave and microwave regions and longer wavelengths.

It is essential to know the characteristics of each basic type of enemy radar so that the electronic counter-measure (ECM) equipment carried by bomber aircraft can be suitably designed and built. The purpose of such ECM instru-ments is to interfere with enemy air-defence radars. The interference may consist of generating spurious signals which create an illusion in enemy radar systems of several bombers, thereby confusing enemy anti-aircraft missiles. Details such as the operating frequency of enemy radar and the speed at which its antennas rotate, the rate at which it transmits successive pulses and the length of time each pulse lasts, are measured. The ECM equipment carried by the bombers must also be designed to cope with enemy electronic counter-counter measures (ECCMs).

These satellites are also used to locate military radio stations and to eaves-drop on military communications.

Satellite orbits

The ideal altitude for electronic reconnaissance satellites is slightly higher than that for photographic reconnaissance satellites. However, the altitude for an electronic reconnaissance satellite should not be so high that its sensitivity is significantly reduced. It is true that the higher the satellite, the longer its life-time, but a high altitude reduces the payload that the launch vehicle can carry. In view of these factors, the most commonly chosen altitude for most electronic reconnaissance satellites is in the range of from 300 to 400 km, at which altitude the satellite's lifetime is in the range of several years. However, the useful life-time is in fact determined by the lifetime of the satellite's batteries and solar cells as well as that of its complex electronic receivers and tape recorders.

The US programme

In addition to the US photographic reconnaissance satellites discussed above, electronic reconnaissance satellites were developed using so-called Elint (elec-tronic intelligence) or ferret satellites. These satellites are launched into orbits with perigee heights of about 300–500 km, that is, higher than those used for photographic reconnaissance, and have considerably longer orbital lives – of the order of years rather than days. As the satellite passes over areas of interest, radar signals and other sources of electromagnetic radiation are recorded on tape. The tapes are then played back and the signals transmitted to ground receiving stations and deciphered. At the end of the mission, the satellites re-enter the Earth's atmosphere and burn up.

As with photographic reconnaissance satellites, two types of ferret satellites may also be in use: one type is used for large-area surveillance, for locating the approximate positions of radars and for determining their frequency bands, and the second type, larger and more complex, is then used to obtain more detailed information on the characteristics of the radars of interest. Often a pair of satellites have been used, one going into a 300–500 km orbit and the second going into a lower orbit with a perigee height of about 200 km. The perigee heights would suggest that the first is a ferret type of satellite and that the second is a recoverable photographic reconnaissance satellite [15]. The ferret satellites in this group are usually octagonal in shape and weigh about 60 kg. The first US ferret satellite of this type was launched on 29 August 1963.

Until 1967, US electronic reconnaissance satellites were launched using Thor-Agena boosters, but from then until 1972 TAT/Agena boosters were used. The payloads placed into orbits have been about 1500 kg. From October 1972, these boosters were no longer used to orbit the electronic satellites; instead the Big Bird satellites carried them into orbit. At a later stage the satellites were ejected from the Big Bird into independent orbits with much greater perigee heights. However, there appear to be no US electronic reconnaissance satellites launched during 1975. Although the Big Bird satellite launched on 4 December 1975 ejected an object into an independent orbit, its orbit was elliptical with a perigee and an apogee of 236 km and 1558 km, respectively, which indicates that it may not have been an electronic reconnaissance satellite. The weight and shape of the satellite are not known.

Ground tracks of a typical US electronic reconnaissance satellite show that the satellite covers the Earth's surface extensively without concentrating over any specific areas. The orbital characteristics of these satellites are listed in Table 3B.1.

The Soviet programme

It is not certain which of the large number of Soviet satellites in the Cosmos series are used for gathering electronic intelligence data. In the *Table of Earth Satellites* published by the Royal Aircraft Establishment, a large number of Soviet satellites, ellipsoidal in shape and weighing about 400 kg, are listed. Of these, those launched from Plesetsk at an orbital inclination of about 71° and with orbital periods of 92 and 95 min have perigee heights of about 300 km or more. Their orbital lives are also considerably longer and so they are therefore possibly electronic reconnaissance satellites. Satellites launched at an inclination of about 74° and with orbital periods of about 95 min may also fall into this group [40].

It is possible that a number of Soviet photographic reconnaissance satellites may also be performing electronic reconnaissance. Certain photographic reconnaissance satellites exhibit a change in telemetry, which is due to the activation of a UHF/VHF transmitter for the rapid replay of stored data which may consist of information collected by electronic sensors [23].

The ground tracks of one of these Soviet satellites, Cosmos 749, calculated for a 14-day period, are plotted in Figure 3.21. It can be seen that, unlike those of the US electronic reconnaissance satellite, the Soviet satellite ground tracks repeat and thus cover only specific regions of the Earth. Therefore, in order to get fuller coverage, a number of satellites with different orbital parameters must be used. In fact, a number of satellites appear to fall in a regular pattern. For example the orbital planes of the satellites with an orbital period of 95 min are spaced at 45° intervals. There are now eight such satellites operating at a time [41]. The ground tracks of the satellites are three degrees apart, giving more complete coverage of the Earth's surface. Orbital characteristics of all the Soviet electronic reconnaissance satellites are given in Table 3B.2.

Figure 3.21. Ground tracks of a Soviet electronic surveillance satellite, Cosmos 749 (1975-62A) for a 14-day period

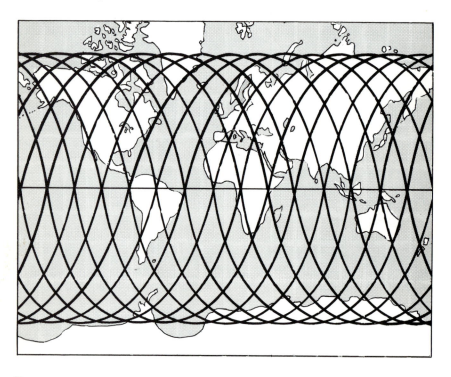

III. Ocean-surveillance satellites

Another development in reconnaissance satellite technology which has recently attracted attention is the development of satellites for ocean surveillance to monitor the location of naval fleets and shore facilities.

The US programme

It is not possible to identify US satellites used only for the purpose of ocean surveillance. However, it is believed that some of the older generation satellites which were launched soon after Big Bird satellites may be carrying infra-red equipment for ocean-surveillance purposes [42]. The infra-red equipment carried by some of the US reconnaissance satellites is similar to that used in the Earth Resources Technology Satellites (Landsat 1 and 2). Photographs taken by cameras on Landsat satellites from a height of 900 km show even such details as small pleasure boats. The resolution improves by a factor of six when photographs are taken from a height of about 150 km, the perigee height of most US photographic reconnaissance satellites. Such infra-red devices are still in the development stage but when they are fully developed it may be possible to detect nuclear submarines travelling at considerable depths [15, 43], since the water used to cool the core of the nuclear reactor of a submarine is expelled into the sea and the temperature of the expelled water may be sufficiently higher than that of the surrounding water that it can be detected by satellite infra-red sensors.

The US Navy (USN) recently launched its first prototype ocean-surveillance satellite into a near-circular orbit using an Atlas rocket. Designed to monitor surface ships, this satellite was built by the US Naval Research Laboratory under the code-name Whitecloud. The satellite contains a number of sensors, including passive infra-red and millimetre-wave radiometers as well as radio-frequency antennas for detecting shipborne radar and communications signals [44]. The satellite also carried three small sub-satellites which are now orbiting the Earth in near-circular orbits similar to that of the main satellite. Each of these sub-satellites is believed to carry an infra-red/millimetre-wave sensor so that, together with the main satellite, a large part of the ocean surface is covered. The sub-satellites are believed to transmit their data to the parent satellite for processing and relay to the Earth stations [45]. The USN is also planning a study of ocean-surveillance satellites equipped with high-resolution radar under a programme called Clipper Bow [46].

The orbital parameters of these satellites are given in Table 3C.1.

The Soviet programme

Launch vehicle type F-1-m has been used to orbit two types of Soviet satellites. Those in the first group belong to the interceptor/destructor satellites which will be described in Chapter 8. A second group are believed to be ocean-surveillance satellites. An important feature of these satellites is that they perform the ocean-surveillance mission while in orbits with perigee and apogee heights of about 250 and 260 km, respectively. After a few weeks, the satellites eject several objects and are then manoeuvred into their new parking orbits at greater perigee and apogee heights of about 870 and 930 km, respectively.

The first of such flights was performed by Cosmos 198, launched on 27 December 1967. In 1970, Cosmos 367 was launched and moved to its higher orbit so rapidly that only the higher orbital parameters were announced [25]. The true nature of these satellites was not learned until 1974 when the US Navy announced that the Soviet Union had been developing an ocean-surveillance satellite system [47].

It is believed that the Soviet satellites are equipped with radar systems and for good resolution, the satellite must fly in a low orbit. The required high power for such radar systems is probably derived from a nuclear power source. If the satellite remained in a low orbit, there would be a risk of contaminating the atmosphere when it decayed. Therefore, it is believed that the satellite is manoeuvred into a higher parking orbit where its lifetime is 600 years [48].

These satellites perform ocean-surveillance missions in pairs: for example, Cosmos 651 and Cosmos 654 were launched in 1974, Cosmos 723 and Cosmos 724 were launched in 1975 and Cosmos 860 and Cosmos 861 were launched in 1976. Orbital characteristics of the Soviet ocean-surveillance satellites are given in Table 3C.2.

IV. Early-warning satellites

In the late 1950s, the warning time of a surprise ICBM attack was about 15 min. In the case of the USA, for example, this was provided by the Ballistic Missile Early-Warning System (BMEWS), a radar system installed in Alaska and Greenland. The range of radars, however, was limited by the curvature of the Earth. The introduction of artificial Earth satellites in the late 1950s altered the situation; enemy missiles can now be detected as soon as they are launched, thereby almost doubling the warning-time. Satellites carrying sensors sensitive to the infra-red radiation emitted by the hot plume of a rocket were developed.

One of the basic problems with satellite-based infra-red sensors was that it was difficult positively to identify an enemy missile. The sensors could not discriminate between radiation from the engine plume and that reflected from high-altitude sunlit clouds. This difficulty was overcome by using infra-red sensors as well as TV cameras on board the early-warning satellites. The resolution of such TV cameras is sufficiently good for surveillance purposes. If the infra-red sensor detects ICBM radiation, a warning signal is transmitted to the ground station. On command from the ground station the satellite TV camera starts to transmit TV pictures to the ground observer who can determine the origin of the infra-red radiation.

Satellite orbits

The ideal orbit for an early-warning satellite is a synchronous equatorial orbit in which the satellite remains in a fixed position relative to the Earth. One problem, however, with such an orbit is that for countries at high latitudes the satellite sensors have a very oblique view. In the early periods this was overcome by orbiting the satellites at an inclination of about 10°. The satellite ground tracks traced a figure eight so that the ground track moves upwards to a latitude of 10°N and then down to 10°S. In this way a more direct view of the extreme regions of the northern and the southern parts of the Earth is obtained. The figure eight repeats once every 24 h. If two satellites are used in such a way that one lags behind the other by 12 h, then more continuous observations of these high-latitude countries can be made.

The US programme

The desire to increase the early-warning time and thus the ability to launch a greater thermonuclear retaliatory attack, as well as advances made in space technology gave impetus to the development of the space-based early-warning system. In 1958 Project MIDAS was established. Under this programme the satellite-based infra-red technique for detecting the launch of an enemy missile was developed. After a series of postponements, an attempt was made on 26 February 1960 to orbit MIDAS 1, using an Atlas/Agena-A booster, but the second stage failed to separate and the launch was a failure. The first successful early-warning research and development (R&D) satellite, MIDAS 2, had a perigee of about 480 km but the operational prototype satellites carrying payloads of 1600–2000 kg were launched into near-circular polar orbits at perigee heights of 3000 km or more, using Atlas/Agena-B or Atlas/Agena-D boosters. The successful launching of MIDAS 2 gave impetus to the development of such photographic reconnaissance satellites as SAMOS. Moreover, it was thought by many at the time that MIDAS provided such a deterrent that it might possibly hold the key to world disarmament.

The operational MIDAS system was expected to consist of six to eight satellites in orbit at all times. Although the satellites were designed to discriminate between the exhaust of a ballistic missile and other heat sources such as large fires and blast furnaces, a problem with the satellite's infra-red sensors was that they could not discriminate between the radiation emitted from the rocket engines and radiation from the Sun, which reached the sensors after reflection from high clouds. A new generation of early-warning satellites was developed, the first of which was launched into a near synchronous equatorial orbit on 6 August 1968. These satellites can probably not only provide early warning of ICBM launches but also detect nuclear explosions. The satellites may also be used for communication purposes [13]. They orbit at very high perigee heights, usually greater than 30 000 km, and have very long orbital lives, greater than one million years.

Early-warning satellites, the BMEWS and the Integrated Missile Early-Warning System (IMEWS), apart from providing early warning of an ICBM attack, also provide a capability to monitor missile tests. The United States launched two early-warning satellites in 1973. It was thought that, in conjunction with over-the-horizon (OTH) radars, these satellites could be used to verify the restrictions on missile testing stipulated in the SALT I ABM Treaty. Both OTH radar and early-warning satellites, when used with other types of radar system, can be used to indicate when and where missile tests are taking place and to provide information on the type of tests being conducted. However, the OTH radars are no longer used.

The orbital characteristics of these US satellites are shown in Table 3D.1. IMEWS 4 was probably launched to supplement IMEWS 2, launched on 5 May 1971. The infra-red sensors on IMEWS 2 are believed to be losing their sensitivity [49] and tests are being carried out to determine the cause. New infra-red sensors are probably also being tested [50]. The US early-warning satellites launched prior to IMEWS 4 were intended only as development models for tests and evaluation.

After a gap of some two years, on 18 June 1975 the United States launched a satellite with orbital characteristics typical of those of early-warning satellites, that is, in an equatorial synchronous orbit. The satellite carries an experimental payload to test a new type of infra-red sensor which will permit more accurate mid-course trajectory tracking of a missile. Such a device might be used in a new generation of early-warning satellites. Another possibility is that it is a prototype satellite of a smaller, low-cost version of the present early-warning satellites [51]. An IMEWS satellite was also launched on 14 December 1975 but it developed some technical difficulty, the cause of which is still not known [52]. The latest satellite was launched on 6 February 1977. Recently the USAF has been considering development of new infra-red sensors for detecting not only ICBMs but also strategic aircraft from space. A satellite, designated P-50B (Teal Ruby), is part of the DoD's Space Test Program and will be orbited by the space shuttle orbiter [53].

The Soviet programme

It is difficult to identify those Soviet satellites which are equivalent to the US MIDAS and early-warning satellites, although the USSR has probably developed similar systems and may be employing different techniques to obtain information similar to that obtained by the MIDAS and other US early-warning satellites. One guess is that Cosmos 159, and some satellites in the Electron series (satellites launched to study the Van Allen radiation belt), could carry sensors similar to those used in the MIDAS series and that Molniya satellites, which spend a long time over North America while still visible from Soviet ground stations, could be performing missile detection missions [54a, 55]. Early-warning satellites are ideally suited for learning of the launch of intercontinental ballistic missiles with nuclear warheads, so it is therefore reasonable to assume that the Soviet Union has also developed such a satellite system. It has been suggested that vehicle type A-2-e has been used to orbit early-warning satellites into 12-h orbits from Plesetsk [25].

It seems that the recently launched Cosmos 775 is probably the first Soviet early-warning satellite in synchronous orbit [56]. This satellite was launched using the type D-1-e vehicle from Tyuratam. It was placed in a synchronous orbit, the plane of which was inclined at 0.03° to the equatorial plane. The perigee and apogee heights of the satellite were 35 737 and 36 220 km, respectively. The satellite was placed into position over the Atlantic Ocean where it could observe US submarine-launched ballistic missiles (SLBMs).

The details of the orbital characteristics of these satellites are shown in Table 3D.2.

V. Nuclear-explosion detection satellites

This is the last of the reconnaissance category of satellites which have been developed to check violations of the treaty banning nuclear explosions in the atmosphere and in outer space. It is relatively simple, in theory, to detect nuclear explosions in space using satellites equipped with radiation detectors. There is a high degree of vacuum in space so that radiation and fission products from a nuclear explosion expand freely, unlike explosions in the Earth's atmosphere. However, the presence in space of Van Allen radiation belts and solar flares complicates the detection of radiation by this technique.

If a nuclear explosion takes place in the radiation belts, radiation caused by the explosion may not be detected by the detectors on board the satellite because there is already a very high level of radiation present in the belts. Therefore, to use satellites effectively to verify the observance of the Partial Test Ban Treaty, other techniques are required. A nuclear explosion produces many other phenomena which may possibly be used for the purpose. It is possible that the satellites carry, besides the radiation detectors or in place of them, optical instruments which would analyse by spectral analysis the emission spectra of the most abundant chemical elements of the nuclear bomb or its fission products (lithium, uranium, plutonium, barium, and so on). This method of detection of explosion in space or in the atmosphere would by no means be hampered by radiation belts. Detectors sensitive to X-rays are also used for detection of nuclear explosions in space.

The US programme

US satellites for the detection of nuclear explosions in the atmosphere and in space have been developed under the Vela programme. Between 1963 and 1965, Vela satellites were launched using Atlas/Agena-D boosters and with payload capacities of 150 kg. During the period 1967–70, the payloads were increased to between 230 and 260 kg and Titan-3C boosters were used. The satellites were orbited at perigee heights of between 100 000 and 111 500 km and therefore have an extremely long orbital life – longer than one million years. They were launched in pairs in near circular orbits, the two satellites occupying virtually the same orbit but diametrically opposed to each other, so that observation could be made simultaneously from opposite sides of the Earth [15].

It was initially planned to use six satellites below the radiation belts (under about 1000 km) and six above the belts, in about 65 000-km orbits. Each satellite would carry Geiger-Mueller and scintillation counters to detect gamma radiation and neutrons produced in a nuclear explosion. The satellite's radio transmitters would then transmit the data collected to their base.

In April 1970 the United States announced that Vela 11 and Vela 12 would be the last of the Vela series. However, interest in verifying the observance of the Partial Test Ban Treaty has not diminished, and the sensors for detecting nuclear tests are now probably carried by the new early-warning system satellites. The characteristics of these are also listed in Table 3D.1.

The Soviet programme

In the case of the Soviet Union it is difficult to identify satellites which perform the nuclear-explosion detection mission. For example, the Cosmos series includes no satellite with orbital characteristics analogous to those of US nuclear-explosion detection satellites. However, it is possible that some satellites in the Electron series could carry sensors similar to those used in the Vela satellites [54b, 55].

Appendix 3A

Tables of photographic reconnaissance satellites

For abbreviations, acronyms and conventions, see page xv.

The designation of each satellite is recognized internationally and is given by the World Warning Agency on behalf of the Committee on Space Research.

More detailed tables of US and Soviet photographic reconnaissance satellites are to be found in *World Armaments and Disarmament, SIPRI Yearbooks 1973* (pp. 76–83, 89–98, 101), *1974* (pp. 299–301), *1975* (pp. 397–99), *1976* (pp. 115–17), and *1977* (pp. 137–38, 154–56). For Chinese photographic reconnaissance satellites, see *SIPRI Yearbooks 1976* (p. 115) and *1977* (p. 175).

Table 3A.1. US photographic reconnaissance satellites

Satellite name and designation	Launch date and time *GMT*	Orbital inclination *deg*	Perigee height *km*	Apogee height *km*	Comments
	1959				
Discoverer 1 (1959-β1)	28 Feb 2150	90	163	968	Doubtful whether it attained orbit
Discoverer 2 (1959-γ1)	13 Apr 2122	90	239	346	Photographic film capsule ejected; lost in Arctic
Discoverer 3 —	3 Jun ..	—	—	—	Failed to orbit
Discoverer 4 —	25 Jun ..	—	—	—	Failed to orbit
Discoverer 5 (1959-ε1)	13 Aug 1858	80	217	739	Film capsule orbited; decayed later
Discoverer 6 (1959-ζ1)	19 Aug 1926	84	212	849	Film capsule ejected; recovery failed
Discoverer 7 (1959-π1)	7 Nov 2024	82	159	847	Film capsule not ejected
Discoverer 8 (1959-λ1)	20 Nov 1926	81	187	1 679	Film capsule overshot recovery area; not recovered
	1960				
Discoverer 9 —	4 Feb ..	—	—	—	Failed to orbit
Discoverer 10 —	19 Feb ..	—	—	—	Failed to orbit
Discoverer 11 (1960-δ1)	15 Apr 2024	80	170	589	Film capsule ejected; recovery failed
Discoverer 12 —	29 Jun ..	—	—	—	Failed to orbit

Satellite name and designation	Launch date and time *GMT*	Orbital inclination *deg*	Perigee height *km*	Apogee height *km*	Comments
Discoverer 13 (1960-θ1)	10 Aug 2038	83	258	683	First ocean film capsule recovery
Discoverer 14 (1960-π1)	18 Aug 1955	80	186	805	First mid-air film capsule recovery
Discoverer 15 (1960-μ1)	13 Sep 2219	81	199	761	Film capsule ejected; lost in sea
SAMOS 1 —	11 Oct ..	—	—	—	Failed to orbit
Discoverer 16 —	26 Oct ..	—	—	—	Failed to orbit
Discoverer 17 (1960-01)	12 Nov 2238	82	190	984	Mid-air film capsule recovery
Discoverer 18 (1960-σ1)	7 Dec 2024	82	243	661	Mid-air film capsule recovery
1961 SAMOS 2 (1961-α1)	31 Jan 2024	97	474	557	Photographs transmitted by radio
Discoverer 20 (1961-ε1)	17 Feb 2024	81	288	786	Film capsule not ejected
Discoverer 22 —	30 Nov ..	—	—	—	Failed to orbit
Discoverer 23 (1961-λ1)	8 Apr 1800	82	200	1 422	Film capsule orbited; decayed later
Discoverer 24	8 Jun	—	—	—	Failed to orbit
Discoverer 25 (1961-ζ1)	16 Jun 2302	82	222	409	Ocean film capsule recovery
Discoverer 26 (1961-π1)	7 Jul 2331	83	228	808	Mid-air film capsule recovery
Discoverer 27 —	21 Jul ..	—	—	—	Failed to orbit
Discoverer 28 —	3 Aug ..	—	—	—	Failed to orbit
Discoverer 29 (1961-χ1)	30 Aug 1932	82	152	542	Ocean film capsule recovery
SAMOS —	9 Sep ..	—	—	—	Failed to orbit
Discoverer 30 (1961-ω1)	12 Sep 1955	83	235	546	Mid-air film capsule recovery
Discoverer 31 (1961-αβ1)	17 Sep 2107	83	235	396	Film capsule not ejected
Discoverer 32 (1961-αγ1)	13 Oct 1926	82	234	395	Mid-air film capsule recovery
Discoverer 33 —	23 Oct ..	—	—	—	Failed to orbit
Discoverer 34 (1961-αε1)	5 Nov 1955	83	227	1 011	Film capsule not ejected

Satellite name and designation	Launch date and time GMT	Orbital inclination deg	Perigee height km	Apogee height km	Comments
Discoverer 35 (1961-αζ1)	15 Nov 2121	82	238	278	Mid-air film capsule recovery
USAF —	22 Nov ..	—	—	—	Failed to orbit
Discoverer 36 (1961-απ1)	12 Dec 2238	81	241	484	Ocean film capsule recovery; photographs also transmitted by radio
USAF (1961-αλ1)	22 Dec 1912	90	244	702	
1962					
Discoverer 37 —	13 Jan ..	—	—	—	Failed to orbit
Discoverer 38 (1962-ε1)	27 Feb 2150	82	208	341	Mid-air film capsule recovery
USAF (1962-η1)	7 Mar 1912	91	251	676	
USAF (1962-λ1)	18 Apr ..	74	200	441	
USAF (1962-π1)	26 Apr 2136	74	..	..	Orbital parameters uncertain
USAF (1962-ρ1)	28 Apr 2248	73	180	475	
USAF (1962-σ1)	15 May 1938	82	305	634	
USAF (1962-φ1)	30 May 0029	74	199	319	
USAF (1962-χ1)	2 Jun 0043	74	211	385	
USAF (1962-τ1)	17 Jun ..	..	..	..	
USAF (1962-αβ1)	23 Jun 0029	76	213	293	
USAF (1962-αγ1)	28 Jun 0112	76	211	689	
USAF (1962-αζ1)	18 Jul 2053	96	184	236	
USAF (1962-αη1)	21 Jul 0058	70	208	381	
USAF (1962-αθ1)	28 Jul 0029	71	225	386	
USAF (1962-απ1)	2 Aug 0029	82	204	418	
USAF (1962-αλ1)	5 Aug 1800	96	205	205	
USAF (1962-ασ1)	29 Aug 0012	65	187	400	
USAF (1962-αγ1)	1 Sep 2238	83	300	609	

Satellite name and designation	Launch date and time GMT	Orbital inclination deg	Perigee height km	Apogee height km	Comments
USAF (1962-αχ1)	17 Sep 2346	82	204	668	Carried Environmental Research Satellite; this not ejected
USAF (1962-ββ1)	29 Sep 2346	65	203	376	
USAF (1962-βε1)	9 Oct 1858	82	213	427	
USAF (1962-β01)	5 Nov 2219	75	208	409	
USAF (1962-βπ1)	11 Nov 2024	96	206	206	
USAF (1962-βρ1)	24 Nov 2205	65	204	337	
USAF (1962-βσ1)	4 Dec · 2136	65.	194	273	
USAF (1962-βφ1)	14 Dec 2122	71	199	392	
	1963				
USAF (1963-02A)	7 Jan 2107	82	205	399	
USAF —	28 Feb ..	—	—	—	Failed to orbit
USAF —	18 Mar ..	—	—	—	Failed to orbit
USAF (1963-07A)	1 Apr 2248	75	201	408	
USAF —	26 Apr ..	—	—	—	Failed to orbit
USAF (1963-16A)	18 May 2234	75	153	497	
USAF (1963-19A)	12 Jun 2400	82	192	419	
USAF (1963-25A)	27 Jun	82	196	396	
USAF (1963-28A)	12 Jul 2238	95	164	164	
USAF (1963-29A)	18 Jul 2400	83	194	387	
USAF (1963-32A)	30 Jul 2400	75	157	411	
USAF (1963-34A)	25 Aug 0029	75	161	320	
USAF (1963-35A)	29 Aug 1912	82	292	324	
USAF (1963-36A)	6 Sep 1926	94	169	263	
USAF (1963-37A)	23 Sep 2248	75	161	441	

Satellite name and designation	Launch date and time GMT	Orbital inclination deg	Perigee height km	Apogee height km	Comments
USAF (1963-41A)	25 Oct 1858	99	144	332	
USAF (1963-42A)	29 Oct 2107	90	279	345	
USAF —	9 Nov ..	—	—	—	Failed to orbit
USAF (1963-48A)	27 Nov 2107	70	175	386	
USAF (1963-51A)	18 Dec 2150	98	122	266	
USAF (1963-55A)	21 Dec 2150	65	176	355	
1964					
USAF (1964-08A)	15 Feb 2136	75	179	444	May carry capsules although radio-transmission type
USAF (1964-09A)	25 Feb 1858	96	173	190	
USAF (1964-12A)	11 Mar 2010	96	163	203	
USAF —	24 Mar ..	—	—	—	Failed to orbit
USAF (1964-20A)	23 Apr 1843	104	150	336	
USAF (1964-22A)	27 Apr 2331	80	178	446	May carry capsules although radio-transmission type
USAF (1964-24A)	19 May 1926	101	141	380	
USAF (1964-27A)	4 Jun 2324	80	149	429	May carry capsules although radio-transmission type
USAF (1964-30A)	13 Jun 1550	115	350	364	Payload included Star Flash experiment
USAF (1964-32A)	19 Jun 2317	85	176	461	May carry capsules although radio-transmission type
USAF (1964-36A)	6 Jul 2150	93	121	346	
USAF (1964-37A)	10 Jul 2317	85	180	461	May carry capsules although radio-transmission type
USAF (1964-43A)	5 Aug 2317	80	182	436	May carry capsules although radio-transmission type
USAF (1964-45A)	14 Aug 2248	96	149	307	
USAF (1964-48A)	21 Aug 1550	115	349	363	Payload included Star Flash experiment
USAF (1964-56A)	14 Sep 2248	85	172	466	May carry capsules although radio-transmission type
USAF (1964-58A)	23 Sep 2010	93	145	303	

Satellite name and designation	Launch date and time GMT	Orbital inclination deg	Perigee height km	Apogee height km	Comments
USAF (1964-61A)	5 Oct 2150	80	182	440	May carry capsules although radio-transmission type
USAF —	8 Oct . .	—	—	—	Failed to orbit
USAF (1964-67A)	17 Oct 2248	76	189	416	May carry capsules although radio-transmission type
USAF (1964-68A)	23 Oct 1829	96	139	271	
USAF (1964-71A)	2 Nov 2136	80	180	448	May carry capsules although radio-transmission type
USAF (1964-75A)	18 Nov 2038	70	180	339	May carry capsules although radio-transmission type
USAF (1964-79A)	4 Dec 1858	97	158	357	
USAF (1964-85A)	19 Dec 2107	75	183	410	May carry capsules although radio-transmission type
USAF (1964-87A)	21 Dec 1912	70	238	264	
1965					
USAF (1965-02A)	15 Jan 2107	75	180	420	May carry capsules although radio-transmission type
USAF (1965-05A)	23 Jan 2010	103	146	291	
USAF (1965-13A)	25 Feb 2150	75	177	377	May carry capsules although radio-transmission type
USAF (1965-19A)	12 Mar 1926	108	155	247	
USAF (1965-26A)	25 Mar 2107	96	186	265	May carry capsules although radio-transmission type
USAF (1965-31A)	28 Apr 2010	96	180	259	
USAF (1965-33A)	29 Apr 2136	85	178	473	May carry capsules although radio-transmission type
USAF (1965-37A)	18 May 1800	75	198	331	May carry capsules although radio-transmission type
USAF (1965-41A)	27 May 1926	96	149	267	
USAF (1965-45A)	9 Jun 2248	75	176	362	May carry capsules although radio-transmission type
USAF (1965-50B)	25 Jun 1926	108	151	283	
USAF —	12 Jul . .	—	—	—	Failed to orbit
USAF (1965-57A)	19 Jul 2248	85	182	464	May carry capsules although radio-transmission type
USAF (1965-62A)	3 Aug 1912	108	149	307	

Satellite name and designation	Launch date and time GMT	Orbital inclination deg	Perigee height km	Apogee height km	Comments
USAF (1965-67A)	17 Aug 2053	70	180	407	May carry capsules although radio-transmission type
USAF —	2 Sep ..	—	—	—	Failed to orbit
USAF (1965-74A)	22 Sep 2122	80	191	364	May carry capsules although radio-transmission type
USAF (1965-76A)	30 Sep 1926	96	158	264	
USAF (1965-79A)	5 Oct 1746	75	203	323	
USAF (1965-86A)	28 Oct 2122	75	176	430	May carry capsules although radio-transmission type
USAF (1965-90A)	8 Nov 1926	94	145	277	Possible electronic reconnaissance mission
USAF (1965-102A)	9 Dec 2107	80	183	437	May carry capsules although radio-transmission type
USAF (1965-110A)	24 Dec 2107	80	178	446	May carry capsules although radio-transmission type
1966					
USAF (1966-02b)	19 Jan 2010	94	154	246	
USAF (1966-07A)	2 Feb 2136	75	185	425	
USAF (1966-12A)	15 Feb 2024	97	148	293	
USAF (1966-18A)	9 Mar 2217	75	178	432	
USAF (1966-22B)	18 Mar 2024	101	152	284	
USAF (1966-29A)	7 Apr 2248	75	193	312	
USAF (1966-32A)	19 Apr 1912	117	145	398	
USAF —	3 May ..	—	—	—	Failed to orbit
USAF (1966-39A)	14 May 1829	111	133	358	
USAF (1966-42A)	24 May 0155	66	179	271	
USAF (1966-48A)	3 Jun 1926	87	143	288	
USAF (1966-55A)	21 Jun 2136	80	194	367	
USAF (1966-62A)	12 Jul 1800	96	137	236	
USAF (1966-69A)	29 Jul 1843	94	158	250	

Satellite name and designation	Launch date and time GMT	Orbital inclination deg	Perigee height km	Apogee height km	Comments
USAF (1966-72A)	8 Aug 2107	100	194	287	
USAF (1966-74A)	16 Aug 1829	93	146	358	
USAF (1966-83A)	16 Sep 1800	94	148	333	
USAF (1966-85A)	20 Sep 2107	85	188	442	
USAF (1966-86A)	28 Sep 1912	94	151	296	
USAF (1966-90A)	12 Oct 1912	91	155	287	
USAF (1966-98A)	2 Nov 2024	91	159	305	
USAF (1966-102A)	8 Nov 1955	100	172	318	
USAF (1966-109A)	5 Dec 2107	105	137	388	
USAF (1966-113A)	14 Dec 1814	110	138	368	
1967					
USAF (1967-02A)	14 Jan 2122	80	180	380	
USAF (1967-07A)	2 Feb 1955	103	136	357	
USAF (1967-15A)	22 Feb 2248	80	180	380	
USAF (1967-16A)	24 Feb 1955	107	135	414	
USAF (1967-29A)	30 Mar 1858	85	167	326	
USAF —	26 Apr ..	—	—	—	Failed to orbit
USAF (1967-43A)	9 May 2150	85	200	777	
USAF (1967-50A)	22 May 1829	92	135	293	
USAF (1967-55A)	4 Jun 1800	105	149	456	
USAF (1967-62A)	16 Jun 2136	80	181	367	
USAF (1967-64A)	20 Jun 1619	111	127	325	
USAF (1967-76A)	7 Aug 2136	80	174	346	
USAF (1967-79A)	16 Aug 1702	112	142	449	

Satellite name and designation	Launch date and time GMT	Orbital inclination deg	Perigee height km	Apogee height km	Comments
USAF (1967-87A)	15 Sep 1938	80	150	389	
USAF (1967-90A)	19 Sep 1829	106	122	401	
USAF (1967-103A)	25 Oct 1912	112	136	429	
USAF (1967-109A)	2 Nov 2136	82	183	410	
USAF (1967-121A)	5 Dec 1843	110	137	430	
USAF (1967-122A)	9 Dec 2219	82	158	237	
1968					
USAF (1968-05A)	18 Jan 1858	112	138	404	
USAF (1968-08A)	24 Jan 2219	82	176	430	
USAF (1968-18A)	13 Mar 1955	100	128	407	
USAF (1968-20A)	14 Mar 2248	83	178	391	
USAF (1968-31A)	17 Apr 1702	112	134	427	
USAF (1968-39A)	1 May 2136	83	164	234	
USAF (1968-47A)	5 Jun 1731	111	123	456	
USAF (1968-52A)	20 Jun 2150	85	193	326	
USAF (1968-64A)	6 Aug 1634	110	142	395	
USAF (1968-65A)	7 Aug 2136	82	152	257	
USAF (1968-74A)	10 Sep 1829	106	125	404	
USAF (1968-78A)	18 Sep 2136	83	167	393	
USAF (1968-98A)	3 Nov 2136	82	150	288	
USAF (1968-99A)	6 Nov 1912	106	130	390	
USAF (1968-108A)	4 Dec 1926	106	136	736	
USAF (1968-112A)	12 Dec 2219	81	169	248	

Satellite name and designation	Launch date and time GMT	Orbital inclination deg	Perigee height km	Apogee height km	Comments
1969					
USAF (1969-07A)	22 Jan 1912	106	142	1 090	Manoeuvrable
USAF (1969-10A)	5 Feb 2248	82	178	238	
USAF (1969-19A)	4 Mar 1926	92	134	461	
USAF (1969-26A)	19 Mar 2136	84	179	241	
USAF (1969-39A)	15 Apr 1731	109	135	410	Manoeuvrable; returned to Earth and recovered intact
USAF (1969-41A)	2 May 0155	65	179	326	
USAF (1969-50A)	3 Jun 1648	110	137	414	Returned to Earth and recovered intact
USAF (1969-63A)	24 Jul 0126	75	178	220	Manoeuvrable
USAF (1969-74A)	22 Aug 1605	108	133	366	Manoeuvrable; returned to Earth and recovered intact
USAF (1969-79A)	22 Sep 2107	85	178	253	
USAF (1969-95A)	24 Oct 1814	108	136	740	
USAF (1969-105A)	4 Dec 2136	82	159	251	
1970					
USAF (1970-02A)	14 Jan 1843	110	134	383	Returned to Earth and recovered intact
USAF (1970-16A)	4 Mar 2219	88	167	257	
USAF (1970-31A)	15 Apr 1550	111	130	388	
USAF (1970-40A)	20 May 2136	83	162	247	
USAF (1970-48A)	25 Jun 1453	109	129	389	
USAF (1970-54A)	23 Jul 0126	60	158	398	
USAF (1970-61A)	18 Aug 1453	111	151	365	Returned to Earth and recovered intact
USAF (1970-90A)	23 Oct 1746	111	135	395	
USAF (1970-98A)	18 Nov 2122	83	185	232	

Satellite name and designation	Launch date and time GMT	Orbital inclination deg	Perigee height km	Apogee height km	Comments
1971					
USAF (1971-05A)	21 Jan 1829	111	139	418	
USAF —	17 Feb ..	—	—	—	Failed to orbit
USAF (1971-22A)	24 Mar 2107	82	157	246	
USAF (1971-33A)	22 Apr 1536	111	132	401	
USAF (1971-56A)	15 Jun 1843	96	184	300	Big Bird
USAF (1971-70A)	12 Aug 1410	111	137	424	
USAF (1971-76A)	10 Sep 2136	75	156	244	
USAF (1971-92A)	23 Oct 1717	111	134	416	
1972					
USAF (1972-02A)	20 Jan 1836	97	156	331	Big Bird
USAF (1972-16A)	17 Mar 1702	111	130	409	
USAF (1972-32A)	19 Apr 2150	82	155	277	
USAF (1972-39A)	25 May 1841	96	158	306	
USAF (1972-52A)	7 Jul 1746	97	174	251	Big Bird
USAF (1972-68A)	1 Sep 1746	111	140	380	
USAF (1972-79A)	10 Oct 1800	97	160	281	Big Bird
USAF (1972-103A)	21 Dec 1746	111	139	378	Manoeuvrable
1973					
USAF (1973-14A)	9 Mar 2038	96	152	270	Big Bird
USAF (1973-28A)	16 May 1634	111	136	352	
USAF —	26 Jun ..	—	—	—	Failed to orbit
USAF (1973-46A)	13 Jul 1955	96	156	269	Big Bird
USAF (1973-68A)	27 Sep 1717	111	131	385	
USAF (1973-88A)	10 Nov 1955	97	159	275	Big Bird

Satellite name and designation	Launch date and time GMT	Orbital inclination deg	Perigee height km	Apogee height km	Comments
1974					
USAF (1974-07A)	13 Feb 1800	110	134	393	
USAF (1974-20A)	10 Apr 0824	95	153	285	Big Bird
USAF (1974-42A)	6 Jun 1634	111	136	394	
USAF (1974-65A)	14 Aug 1550	111	135	402	
USAF (1974-85A)	29 Oct 1925	97	162	271	Big Bird
1975					
USAF (1975-32A)	18 Apr 1648	111	134	401	
USAF (1975-51A)	8 Jun 1829	96	154	269	Big Bird
USAF (1975-98A)	9 Oct 1912	96	125	356	
USAF (1975-114A)	4 Dec 2038	96	157	234	Big Bird
1976					
USAF (1976-27A)	22 Mar 1814	96	125	347	
USAF (1976-65A)	8 Jul 1843	97	159	242	Big Bird
USAF (1976-94A)	15 Sep 1858	96	135	330	
USAF (1976-125A)	19 Dec 1829	97	247	533	Big Bird

Table 3A.2. Soviet photographic reconnaissance satellites

Satellite name and designation	Launch date and time GMT	Orbital inclination deg	Perigee height km	Apogee height km	Comments
1962					
Cosmos 4 (1962-ξ1)	26 Apr 1005	65	282	317	First generation; low-resolution cameras
Cosmos 7 (1962-αζ1)	28 Jul 0922	65	197	356	First generation; low-resolution cameras
Cosmos 9 (1962-αω1)	27 Sep 0936	65	292	346	First generation; low-resolution cameras

Satellite name and designation	Launch date and time GMT	Orbital inclination deg	Perigee height km	Apogee height km	Comments
Cosmos 10 (1962-βξ1)	17 Oct 0922	65	197	367	First generation; low-resolution cameras
Cosmos 12 (1962-βω1)	22 Dec 0922	65	198	392	First generation; low-resolution cameras
1963					
Cosmos 13 (1963-06A)	21 Mar 0824	65	192	324	First generation; low-resolution cameras
Cosmos 15 (1963-11A)	22 Apr 0824	65	160	358	First generation; low-resolution cameras
Cosmos 16 (1963-12A)	28 Apr 0936	65	194	388	First generation; low-resolution cameras
Cosmos 18 (1963-18A)	24 May 1048	65	196	288	First generation; low-resolution cameras
Cosmos 20 (1963-40B)	18 Oct 0936	65	205	302	First generation; low-resolution cameras
Cosmos 22 (1963-45A)	16 Nov 1048	65	192	381	Second generation; high-resolution cameras
Cosmos 24 (1963-52A)	19 Dec 0922	65	204	391	First generation; low-resolution cameras
1964					
Cosmos 28 (1964-17A)	4 Apr 0936	65	213	373	First generation; low-resolution cameras
Cosmos 29 (1964-21A)	25 Apr 1019	65	203	296	First generation; low-resolution cameras
Cosmos 30 (1964-23A)	18 May 0950	65	206	366	Second generation; high-resolution cameras
Cosmos 32 (1964-29A)	10 Jun 1048	51	213	319	First generation; low-resolution cameras
Cosmos 33 (1964-33A)	23 Jun 1019	65	209	293	First generation; low-resolution cameras
Cosmos 34 (1964-34A)	1 Jul 1117	65	202	348	Second generation; high-resolution cameras
Cosmos 35 (1964-39A)	15 Jul 1131	51	218	258	First generation; low-resolution cameras
Cosmos 37 (1964-44A)	14 Aug 0936	65	207	287	First generation; low-resolution cameras
Cosmos 45 (1964-55A)	13 Sep 0950	65	207	313	Second generation; high-resolution cameras
Cosmos 46 (1964-59A)	24 Sep 1200	51	211	264	First generation; low-resolution cameras
Cosmos 48 (1964-66A)	14 Oct 0950	65	204	284	First generation; low-resolution cameras
Cosmos 50 (1964-70A)	28 Oct 1048	51	190	230	First generation; low-resolution cameras; exploded in orbit

Satellite name and designation	Launch date and time GMT	Orbital inclination deg	Perigee height km	Apogee height km	Comments
1965					
Cosmos 52 (1965-01A)	11 Jan 0936	65	203	298	First generation; low-resolution cameras
Cosmos 59 (1965-15A)	7 Mar 0907	65	217	310	Second generation; high-resolution cameras
Cosmos 64 (1965-25A)	25 Mar 1005	65	201	267	First generation; low-resolution cameras
Cosmos 65 (1965-29A)	17 Apr 0950	65	207	319	Second generation; high-resolution cameras
Cosmos 66 (1965-35A)	7 May 0950	65	202	282	First generation; low-resolution cameras
Cosmos 67 (1965-40A)	25 May 1048	52	200	346	Second generation; high-resolution cameras
Cosmos 68 (1965-46A)	15 Jun 1005	65	209	315	First generation; low-resolution cameras
Cosmos 69 (1965-49A)	25 Jun 0950	65	212	305	Second generation; high-resolution cameras
Cosmos 77 (1965-61A)	3 Aug 1102	52	201	280	Second generation; high-resolution cameras
Cosmos 78 (1965-66A)	14 Aug 1117	69	218	298	First generation; low-resolution cameras
Cosmos 79 (1965-69A)	25 Aug 1019	65	205	338	Second generation; high-resolution cameras
Cosmos 85 (1965-71A)	9 Sep 0936	65	204	297	Second generation; high-resolution cameras
Cosmos 91 (1965-75A)	23 Sep 0907	65	204	324	Second generation; high-resolution cameras
Cosmos 92 (1965-83A)	16 Oct 0810	65	201	334	Second generation; high-resolution cameras
Cosmos 94 (1965-85A)	28 Oct 0824	65	205	271	Second generation; high-resolution cameras
Cosmos 98 (1965-97A)	27 Nov 0824	65	205	547	First generation; low-resolution cameras
Cosmos 99 (1965-103A)	10 Dec 0810	65	203	309	First generation; low-resolution cameras
1966					
Cosmos 104 (1966-01A)	7 Jan 0824	65	193	380	First generation; low-resolution cameras
Cosmos 105 (1966-03A)	22 Jan 0838	65	204	310	First generation; low-resolution cameras
Cosmos 107 (1966-10A)	10 Feb 0853	65	204	310	First generation; low-resolution cameras
Cosmos 109 (1966-14A)	19 Feb 0853	65	202	295	Second generation; high-resolution cameras
Cosmos 112 (1966-21A)	17 Mar 1033	72	207	545	First generation; low-resolution cameras

Satellite name and designation	Launch date and time GMT	Orbital inclination deg	Perigee height km	Apogee height km	Comments
Cosmos 113 (1966-23A)	21 Mar 0936	65	207	313	Second generation; high-resolution cameras
Cosmos 114 (1966-28A)	6 Apr 1146	73	210	343	Second generation; high-resolution cameras
Cosmos 115 (1966-33A)	20 Apr 1048	65	201	294	First generation; low-resolution cameras
Cosmos 117 (1966-37A)	6 May 1102	65	205	298	First generation; low-resolution cameras
Cosmos 120 (1966-50A)	8 Jun 1102	52	205	285	Second generation; low-resolution cameras
Cosmos 121 (1965-54A)	17 Jun 1102	73	200	333	Second generation; high-resolution cameras
Cosmos 124 (1966-64A)	14 Jul 1033	52	205	286	Second generation; low-resolution cameras
Cosmos 126 (1966-68A)	28 Jul 1048	52	204	350	Second generation; high-resolution cameras
Cosmos 127 (1966-71A)	8 Aug 1117	52	201	267	Second generation; high-resolution cameras
Cosmos 128 (1966-79A)	27 Aug 0950	65	213	319	Second generation; high-resolution cameras
Cosmos 129 (1966-91A)	14 Oct 1214	65	180	312	First generation; low-resolution cameras
Cosmos 130 (1966-93A)	20 Oct 0853	65	208	314	Second generation; high-resolution cameras
Cosmos 131 (1966-105A)	12 Nov 0950	73	204	337	Second generation; high-resolution cameras
Cosmos 132 (1966-106A)	19 Nov 0810	65	210	276	First generation; low-resolution cameras
Cosmos 134 (1966-108A)	3 Dec 0810	65	201	294	Second generation; high-resolution cameras
Cosmos 136 (1966-115A)	19 Dec 1200	65	188	280	First generation; low-resolution cameras
1967					
Cosmos 138 (1967-04A)	19 Jan 1243	65	191	273	First generation; low-resolution cameras
Cosmos 141 (1967-12A)	8 Feb 1019	73	205	316	Second generation; high-resolution cameras
Cosmos 143 (1967-17A)	27 Feb 0824	65	204	297	First generation; low-resolution cameras
Cosmos 147 (1967-22A)	13 Mar 1214	65	195	301	First generation; low-resolution cameras
Cosmos 150 (1967-25A)	22 Mar 1243	66	204	350	Second generation; high-resolution cameras
Cosmos 153 (1967-30A)	4 Apr 1355	65	199	279	First generation; low-resolution cameras

Satellite name and designation	Launch date and time GMT	Orbital inclination deg	Perigee height km	Apogee height km	Comments
Cosmos 155 (1967-33A)	12 Apr 1102	52	193	272	Second generation; high-resolution cameras
Cosmos 157 (1967-44A)	12 May 1033	51	249	262	First generation; low-resolution cameras
Cosmos 161 (1967-49A)	22 May 1355	66	201	321	Second generation; high-resolution cameras
Cosmos 162 (1967-54A)	1 Jun 1048	52	196	275	Second generation; low-resolution cameras
Cosmos 164 (1967-57A)	8 Jun 1312	66	185	317	Second generation; low-resolution cameras
Cosmos 168 (1967-67A)	4 Jul 0600	52	198	264	Second generation; low-resolution cameras
Cosmos 172 (1967-78A)	9 Aug 0546	52	200	293	Second generation; high-resolution cameras
Cosmos 175 (1967-85A)	11 Sep 1033	73	211	358	Second generation; high-resolution cameras
Cosmos 177 (1967-88A)	16 Sep 0600	52	200	280	Second generation; low-resolution cameras
Cosmos 180 (1967-93A)	26 Sep 1019	73	208	341	Second generation; low-resolution cameras
Cosmos 181 (1967-97A)	11 Oct 1131	66	194	327	Second generation; low-resolution cameras
Cosmos 182 (1967-98A)	16 Oct 0755	65	210	330	Second generation; high-resolution cameras
Cosmos 190 (1967-110A)	3 Nov 1117	66	191	338	Second generation; high-resolution cameras
Cosmos 193 (1967-117A)	25 Nov 1131	66	202	335	Second generation; low-resolution cameras
Cosmos 194 (1967-119A)	3 Dec 1200	66	201	307	Second generation; high-resolution cameras
Cosmos 195 (1967-124A)	16 Dec 1200	66	207	353	Second generation; low-resolution cameras
1968					
Cosmos 199 (1968-03A)	16 Jan 1200	66	204	364	Second generation; low-resolution cameras; exploded in orbit
Cosmos 201 (1968-09A)	6 Feb 0755	65	204	337	Second generation; high-resolution cameras
Cosmos 205 (1968-16A)	5 Mar 1229	66	199	292	Second generation; low-resolution cameras
Cosmos 207 (1968-21A)	16 Mar 1229	66	201	321	Second generation; high-resolution cameras
Cosmos 208 (1968-22A)	21 Mar 0950	66	208	274	Third generation; low-resolution cameras; emit PDM[a] signals; ejected capsule
Cosmos 210 (1968-24A)	3 Apr 1102	81	200	373	Second generation; low-resolution cameras

Satellite name and designation	Launch date and time GMT	Orbital inclination deg	Perigee height km	Apogee height km	Comments
Cosmos 214 (1968-32A)	18 Apr 1033	81	200	373	Second generation; high-resolution cameras
Cosmos 216 (1968-34A)	20 Apr 1033	52	201	267	Second generation; low-resolution cameras
Cosmos 223 (1968-45A)	1 Jun 1102	73	200	333	Second generation; low-resolution cameras
Cosmos 224 (1968-46A)	4 Jun 0643	52	203	256	Second generation; high-resolution cameras
Cosmos 227 (1968-51A)	18 Jun 0614	52	190	269	Second generation; high-resolution cameras
Cosmos 228 (1968-53A)	21 Jun 1200	52	199	252	Third generation; low-resolution cameras; emit PDM[a] signals; ejected capsule
Cosmos 229 (1968-54A)	26 Jun 1102	73	207	327	Second generation; high-resolution cameras
Cosmos 231 (1968-58A)	10 Jul 1955	65	199	345	Second generation; low-resolution cameras
Cosmos 232 (1968-60A)	16 Jul 1312	65	189	348	Second generation; high-resolution cameras
Cosmos 234 (1968-62A)	30 Jul 0658	52	208	288	Second generation; high-resolution cameras
Cosmos 235 (1968-67A)	9 Aug 0658	52	201	281	Second generation; low-resolution cameras
Cosmos 237 (1968-71A)	27 Aug 1229	65	200	320	Second generation; high-resolution cameras
Cosmos 239 (1968-73A)	5 Sep 0658	52	203	269	Second generation; high-resolution cameras
Cosmos 240 (1968-75A)	14 Sep 0643	52	202	282	Second generation; low-resolution cameras
Cosmos 241 (1968-77A)	16 Sep 1229	65	202	322	Second generation; high-resolution cameras
Cosmos 243 (1968-80A)	23 Sep 0741	71	213	293	Third generation; low-resolution cameras; emit PDM[a] signals; ejected capsule
Cosmos 246 (1968-87A)	7 Oct 1214	65	149	321	Second generation; high-resolution cameras
Cosmos 247 (1968-88A)	11 Oct 1200	65	199	345	Second generation; low-resolution cameras
Cosmos 251 (1968-96A)	31 Oct 0907	65	201	250	Third generation; high-resolution cameras; manoeuvrable; emit morse code signals; ejected capsule
Cosmos 253 (1968-102A)	13 Nov 1200	65	200	333	Second generation; low-resolution cameras
Cosmos 254 (1968-104A)	21 Nov 1214	65	197	335	Second generation; high-resolution cameras
Cosmos 255 (1968-105A)	29 Nov 1243	65	197	317	Second generation; low-resolution cameras
Cosmos 258 (1968-111A)	10 Dec 0824	65	205	298	Second generation; low-resolution cameras

Satellite name and designation	Launch date and time *GMT*	Orbital inclination *deg*	Perigee height *km*	Apogee height *km*	Comments
1969					
Cosmos 263 (1969-03A)	12 Jan 1214	65	200	325	Second generation; low-resolution cameras
Cosmos 264 (1969-08A)	23 Jan 0922	70	209	295	Third generation; high-resolution cameras; manoeuvrable; emit morse code signals
Cosmos 266 (1969-15A)	25 Feb 1019	73	202	336	Second generation; low-resolution cameras
Cosmos 267 (1969-17A)	26 Feb 0824	65	205	329	Second generation; high-resolution cameras
Cosmos 270 (1969-22A)	6 Mar 1214	65	200	331	Second generation; high-resolution cameras
Cosmos 271 (1969-23A)	15 Mar 1214	65	196	324	Second generation; high-resolution cameras
Cosmos 273 (1969-27A)	22 Mar 1214	65	198	329	Second generation; low-resolution cameras
Cosmos 274 (1969-28A)	24 Mar 1005	65	206	300	Second generation; high-resolution cameras
Cosmos 276 (1969-32A)	4 Apr 1019	81	200	371	Second generation; high-resolution cameras
Cosmos 278 (1969-34A)	9 Apr 1258	65	198	310	Second generation; low-resolution cameras
Cosmos 279 (1969-38A)	15 Apr 0824	52	192	267	Second generation; high-resolution cameras
Cosmos 280 (1969-40A)	23 Apr 1005	52	207	250	Third generation; high-resolution cameras; manoeuvrable; emit morse code signals; ejected capsule
Cosmos 281 (1969-42A)	13 May 0922	65	191	303	Second generation; low-resolution cameras
Cosmos 282 (1969-44A)	20 May 0838	65	202	321	Second generation; high-resolution cameras
Cosmos 284 (1969-48A)	29 May 0658	52	205	294	Second generation; high-resolution cameras
Cosmos 286 (1969-52A)	15 Jun 0907	65	200	327	Second generation; high-resolution cameras
Cosmos 287 (1969-54A)	24 Jun 0658	52	188	264	Second generation; low-resolution cameras
Cosmos 288 (1969-55A)	27 Jun 0712	52	199	273	Second generation; high-resolution cameras
Cosmos 289 (1969-57A)	10 Jul 0907	65	194	325	Second generation; high-resolution cameras
Cosmos 290 (1969-60A)	22 Jul 1229	65	194	332	Second generation; low-resolution cameras
Cosmos 293 (1969-71A)	16 Aug 1200	52	208	256	Third generation; low-resolution cameras; emit **PDM**^{a} signals
Cosmos 294 (1969-72A)	19 Aug 1258	65	200	329	Second generation; high-resolution cameras
Cosmos 296 (1969-75A)	29 Aug 0907	65	207	302	Second generation; high-resolution cameras

Satellite name and designation	Launch date and time GMT	Orbital inclination deg	Perigee height km	Apogee height km	Comments
Cosmos 297 (1969-76A)	2 Sep 1102	73	205	309	Second generation; high-resolution cameras
Cosmos 299 (1969-78A)	18 Sep	65	207	284	Second generation; high-resolution cameras
Cosmos 301 (1969-81A)	24 Sep 1214	65	195	289	Second generation; low-resolution cameras
Cosmos 302 (1969-89A)	17 Oct 1146	65	198	321	Second generation; high-resolution cameras
Cosmos 306 (1969-93A)	24 Oct 0950	65	215	299	Third generation; low-resolution cameras; emit PDM^a signals
Cosmos 309 (1969-98A)	12 Nov 1131	65	185	364	Third generation; low-resolution cameras; emit PDM^a signals
Cosmos 310 (1969-100A)	15 Nov 0838	65	204	336	Second generation; high-resolution cameras
Cosmos 313 (1969-104A)	3 Dec 1326	65	198	259	Third generation; low-resolution cameras; emit PDM^a signals
Cosmos 317 (1969-109A)	23 Dec 1355	65	205	280	Third generation; high-resolution cameras; two-tone telemetry transmission; manoeuvrable
1970					
Cosmos 318 (1970-01A)	9 Jan 0922	65	203	277	Third generation; low-resolution cameras; emit PDM^a signals
Cosmos 322 (1970-07A)	21 Jan 1200	65	195	319	Second generation; high-resolution cameras
Cosmos 323 (1970-10A)	10 Feb 1200	65	201	314	Second generation; high-resolution cameras
Cosmos 325 (1970-15A)	4 Mar 1214	65	200	327	Second generation; low-resolution cameras
Cosmos 326 (1970-18A)	13 Mar 0810	81	203	363	Second generation; low-resolution cameras
Cosmos 328 (1970-22A)	27 Mar 1146	73	203	299	Third generation; high-resolution cameras; manoeuvrable; emit morse code signals
Cosmos 329 (1970-23A)	3 Apr 0838	81	198	228	Third generation; low-resolution cameras; emit PDM^a signals
Cosmos 331 (1970-26A)	8 Apr 1019	65	206	320	Second generation; high-resolution cameras
Cosmos 333 (1970-30A)	15 Apr 0907	81	219	239	Third generation; high-resolution cameras; manoeuvrable; emit morse code signals; ejected capsule
Cosmos 344 (1970-38A)	12 May 1019	73	202	329	Second generation; low-resolution cameras
Cosmos 345 (1970-39A)	20 May 0922	52	192	270	Second generation; high-resolution cameras
Cosmos 346 (1970-42A)	10 Jun 0936	52	197	274	Second generation; high-resolution cameras

Satellite name and designation	Launch date and time GMT	Orbital inclination deg	Perigee height km	Apogee height km	Comments
Cosmos 349 (1970-45A)	17 Jun 1258	65	199	332	Second generation; high-resolution cameras
Cosmos 350 (1970-50A)	26 Jun 1200	52	202	258	Third generation; low-resolution cameras; emit PDM[a] signals
Cosmos 352 (1970-52A)	7 Jul 1033	52	207	294	Second generation; high-resolution cameras
Cosmos 353 (1970-53A)	9 Jul 1341	65	204	284	Third generation; low-resolution cameras; emit PDM[a] signals
Cosmos 355 (1970-58A)	7 Aug 0936	65	199	322	Second generation; high-resolution cameras
Cosmos 360 (1970-68A)	29 Aug 0838	65	209	305	Third generation; high-resolution cameras; manoeuvrable; emit morse code signals; ejected capsule
Cosmos 361 (1970-71A)	8 Sep 1033	73	209	298	Third generation; high-resolution cameras; manoeuvrable; emit morse code signals; ejected capsule
Cosmos 363 (1970-74A)	17 Sep 0824	65	208	294	Third generation; low-resolution cameras; emit PDM[a] signals
Cosmos 364 (1970-75A)	22 Sep 1258	65	202	297	Third generation; high-resolution cameras; two-tone telemetry transmission; manoeuvrable; ejected capsule
Cosmos 366 (1970-78A)	1 Oct 0824	65	204	295	Third generation; low-resolution cameras; emit PDM[a] signals
Cosmos 368 (1970-80A)	8 Oct 1243	65	204	400	Third generation; low-resolution cameras; emit PDM[a] signals; ejected capsule
Cosmos 370 (1970-82A)	9 Oct 1102	65	202	288	Third generation; high-resolution cameras; manoeuvrable; emit morse code signals; ejected capsule
Cosmos 376 (1970-92A)	30 Oct 1326	65	207	286	Third generation; high-resolution cameras; manoeuvrable; emit morse code signals; ejected capsule
Cosmos 377 (1970-96A)	11 Nov 0922	65	204	286	Third generation; low-resolution cameras; emit PDM[a] signals
Cosmos 383 (1970-104A)	3 Dec 1355	65	204	279	Third generation; high-resolution cameras; two-tone telemetry transmission; manoeuvrable
Cosmos 384 (1970-105A)	10 Dec 1117	73	203	292	Third generation; low-resolution cameras; emit PDM[a] signals; ejected capsule
Cosmos 386 (1970-110A)	15 Dec 1005	65	215	276	Third generation; high-resolution cameras; manoeuvrable; emit morse code signals; ejected capsule
1971					
Cosmos 390 (1971-01A)	12 Jan 0936	65	204	275	Unclassified

Satellite name and designation	Launch date and time GMT	Orbital inclination deg	Perigee height km	Apogee height km	Comments
Cosmos 392 (1971-04A)	21 Jan 0838	65	204	278	Third generation; low-resolution cameras; emit PDM[a] signals
Cosmos 396 (1971-14A)	18 Feb 1410	65	205	286	Third generation; high-resolution cameras; manoeuvrable; emit morse code signals; ejected capsule
Cosmos 399 (1971-17A)	3 Mar 0936	65	201	283	Third generation; high-resolution cameras; manoeuvrable; emit morse code signals; ejected capsule
Cosmos 401 (1971-23A)	27 Mar 1102	73	185	290	Third generation; high-resolution cameras; manoeuvrable; emit morse code signals; ejected capsule
Cosmos 403 (1971-26A)	2 Apr 0824	81	214	230	Third generation; low-resolution cameras; emit PDM[a] signals
Cosmos 406 (1971-29A)	14 Apr 0810	81	217	246	Third generation; high-resolution cameras; manoeuvrable; emit morse code signals; ejected capsule
Cosmos 410 (1971-40A)	6 May 0629	65	205	280	Third generation; low-resolution cameras; emit PDM[a] signals
Cosmos 420 (1971-43A)	18 May 0810	52	199	257	Third generation; high-resolution cameras; manoeuvrable; emit morse code signals; ejected capsule
Cosmos 424 (1971-48A)	28 May 1033	65	204	282	Manoeuvrable; unclassified; ejected capsule
Cosmos 427 (1971-55A)	11 Jun 1005	73	204	314	Third generation; high-resolution cameras; two-tone telemetry transmission; manoeuvrable; ejected capsule
Cosmos 428 (1971-57A)	24 Jun 0810	52	206	257	Third generation; low-resolution cameras; emit PDM[a] signals
Cosmos 429 (1971-61A)	20 Jul 1005	52	202	252	Third generation; high-resolution cameras; manoeuvrable; emit morse code signals; ejected capsule
Cosmos 430 (1971-62A)	23 Jul 1102	65	199	305	Third generation; high-resolution cameras; manoeuvrable; emit morse code signals; ejected capsule
Cosmos 431 (1971-65A)	30 Jul 0838	52	194	257	Third generation; low-resolution cameras; emit PDM[a] signals
Cosmos 432 (1971-66A)	5 Aug 1005	52	194	259	Third generation; high-resolution cameras; manoeuvrable; emit morse code signals; ejected capsule
Cosmos 438 (1971-77A)	14 Sep 1258	65	208	296	Third generation; high-resolution cameras; two-tone telemetry transmission; manoeuvrable; ejected capsule
Cosmos 439 (1971-78A)	21 Sep 1200	65	207	284	Third generation; low-resolution cameras; emit PDM[a] signals
Cosmos 441 (1971-81A)	28 Sep 0741	65	204	268	Manoeuvrable; unclassified; ejected capsule

Satellite name and designation	Launch date and time GMT	Orbital inclination deg	Perigee height km	Apogee height km	Comments
Cosmos 442 (1971-84A)	29 Sep 1131	73	182	313	Unclassified; ejected capsule
Cosmos 443 (1971-85A)	7 Oct 1229	65	204	301	Third generation; low-resolution cameras; emit PDM[a] signals; ejected capsule
Cosmos 452 (1971-88A)	14 Oct 0907	65	198	260	Manoeuvrable; unclassified; ejected capsule
Cosmos 454 (1971-94A)	2 Nov 1424	65	203	262	Manoeuvrable; unclassified; ejected capsule
Cosmos 456 (1971-98A)	19 Nov 1200	73	178	304	Third generation; high-resolution cameras; manoeuvrable; emit morse code signals; ejected capsule
Cosmos 463 (1971-107A)	6 Dec 0950	65	202	273	Third generation; high-resolution cameras; manoeuvrable; emit morse code signals; ejected capsule
Cosmos 464 (1971-108A)	10 Dec 1102	73	206	375	Manoeuvrable; unclassified; ejected capsule
Cosmos 466 (1971-112A)	16 Dec 0950	65	209	280	Third generation; high-resolution cameras; manoeuvrable; emit morse code signals; ejected capsule
Cosmos 470 (1971-118A)	27 Dec 1410	65	194	260	Third generation; low-resolution cameras; two-tone telemetry transmission; ejected capsule
1972					
Cosmos 471 (1972-01A)	12 Jan 1005	65	201	317	Third generation; high-resolution cameras; manoeuvrable; emit morse code signals; ejected capsule
Cosmos 473 (1972-06A)	3 Feb 0845	65	205	314	Third generation; low-resolution cameras; emit PDM[a] signals
Cosmos 474 (1972-08A)	16 Feb 0936	65	213	317	Third generation; high-resolution cameras; manoeuvrable; emit morse code signals; ejected capsule
Cosmos 477 (1972-13A)	4 Mar 1005	73	202	306	Third generation; low-resolution cameras; emit PDM[a] signals
Cosmos 478 (1972-15A)	15 Mar 1300	65	204	295	Third generation; high-resolution cameras; manoeuvrable; emit morse code signals; ejected capsule
Cosmos 483 (1972-24A)	3 Apr 1020	73	209	313	Third generation; high-resolution cameras; manoeuvrable; emit morse code signals; ejected capsule
Cosmos 484 (1972-26A)	6 Apr 0805	81	196	224	Third generation; low-resolution cameras; emit PDM[a] signals
Intercosmos[b] (1972-27A)	7 Apr 1005	52	203	248	
Cosmos 486 (1972-30A)	14 Apr 0805	81	178	234	Third generation; high-resolution cameras; manoeuvrable; emit morse code signals; ejected capsule

Satellite name and designation	Launch date and time *GMT*	Orbital inclination *deg*	Perigee height *km*	Apogee height *km*	Comments
Cosmos 488 (1972-34A)	5 May 1133	65	207	294	Third generation; high-resolution cameras; two-tone telemetry transmission; ejected capsule
Cosmos 490 (1972-36A)	17 May 1025	65	205	285	Third generation; low-resolution cameras; emit PDM*a* signals
Cosmos 491 (1972-38A)	25 May 0640	65	177	269	Third generation; high-resolution cameras; manoeuvrable; emit morse code signals; ejected capsule
Cosmos 492 (1972-40A)	9 Jun 0712	65	205	323	Third generation; high-resolution cameras; manoeuvrable; emit morse code signals; ejected capsule
Cosmos 493 (1972-42A)	21 Jun 0635	65	203	274	Third generation; low-resolution cameras; manoeuvrable; emit PDM*a* signals
Cosmos 495 (1972-44A)	23 Jun 1126	65	202	278	Unclassified; ejected capsule
Cosmos 499 (1972-51A)	6 Jul 1048	52	204	283	Third generation; high-resolution cameras; manoeuvrable; emit morse code signals; ejected capsule
Cosmos 502 (1972-55A)	13 Jul 1424	65	204	264	Third generation; low-resolution cameras; two-tone telemetry transmission; ejected capsule
Cosmos 503 (1972-56A)	19 Jul 1355	65	202	288	Third generation; high-resolution cameras; manoeuvrable; emit morse code signals; ejected capsule
Cosmos 512 (1972-59A)	28 Jul 1019	65	203	273	Third generation; low-resolution cameras; emit PDM*a* signals
Cosmos 513 (1972-60A)	2 Aug 0824	65	203	320	Third generation; high-resolution cameras; manoeuvrable; emit morse code signals; ejected capsule
Cosmos 515 (1972-63A)	18 Aug 1005	73	189	325	Third generation; high-resolution cameras; two-tone telemetry transmission; manoeuvrable; ejected capsule
Cosmos 517 (1972-67A)	30 Aug 0824	65	204	288	Third generation; low-resolution cameras; emit PDM*a* signals
Cosmos 518 (1972-70A)	15 Sep 0936	73	204	307	Third generation; low-resolution cameras; emit PDM*a* signals; ejected capsule
Cosmos 519 (1972-71A)	16 Sep 0824	71	207	360	Third generation; high-resolution cameras; manoeuvrable; emit morse code signals; ejected capsule
Cosmos 522 (1972-77A)	4 Oct 1200	73	206	316	Third generation; high-resolution cameras; manoeuvrable; emit morse code signals; ejected capsule
Cosmos 525 (1972-83A)	18 Oct 1200	65	207	269	Third generation; low-resolution cameras; emit PDM*a* signals; ejected capsule

Satellite name and designation	Launch date and time GMT	Orbital inclination deg	Perigee height km	Apogee height km	Comments
Cosmos 527 (1972-86A)	31 Oct 1341	65	207	306	Third generation; high-resolution cameras; two-tone telemetry transmission; manoeuvrable; ejected capsule
Cosmos 537 (1972-93A)	25 Nov 0907	65	204	305	Third generation; low-resolution cameras; emit PDM^a signals
Cosmos 538 (1972-99A)	14 Dec 1355	65	205	283	Third generation; high-resolution cameras; manoeuvrable; emit morse code signals; ejected capsule
Cosmos 541 (1972-105A)	27 Dec 1033	81	221	346	Third generation; low-resolution cameras; two-tone telemetry transmission; ejected capsule

1973

Satellite name and designation	Launch date and time GMT	Orbital inclination deg	Perigee height km	Apogee height km	Comments
Cosmos 543 (1973-02A)	11 Jan 1005	65	203	309	Third generation; high-resolution cameras; manoeuvrable; emit morse code signals; ejected capsule
Cosmos 547 (1973-06A)	1 Feb 0838	65	203	310	Third generation; low-resolution cameras; emit PDM^a signals
Cosmos 548 (1973-08A)	8 Feb 1326	65	205	300	Third generation; high-resolution cameras; manoeuvrable; emit morse code signals; ejected capsule
Cosmos 550 (1973-11A)	1 Mar 1243	65	206	317	Third generation; high-resolution cameras; two-tone telemetry transmission; manoeuvrable; ejected capsule
Cosmos 551 (1973-12A)	6 Mar 0922	65	206	296	Third generation; high-resolution cameras; manoeuvrable; emit morse code signals; ejected capsule
Cosmos 552 (1973-16A)	22 Mar 1005	73	204	312	Third generation; low-resolution cameras; emit PDM^a signals; ejected capsule
Cosmos 554 (1973-21A)	19 Apr 0907	73	194	304	Third generation; high-resolution cameras; two-tone telemetry transmission; manoeuvrable; ejected capsule; destroyed in orbit
Cosmos 555 (1973-24A)	25 Apr 1048	81	216	233	Third generation; low-resolution cameras; emit PDM^a signals; ejected capsule
Cosmos 556 (1973-25A)	5 May 0658	81	218	225	Third generation; high-resolution cameras; two-tone telemetry transmission; manoeuvrable; ejected capsule
Cosmos 559 (1973-30A)	18 May 1102	65	204	325	Third generation; high-resolution cameras; two-tone telemetry transmission; manoeuvrable; ejected capsule

Satellite name and designation	Launch date and time GMT	Orbital inclination deg	Perigee height km	Apogee height km	Comments
Cosmos 560 (1973-31A)	23 May 1033	73	203	314	Third generation; high-resolution cameras; manoeuvrable; emit morse code signals; ejected capsule
Cosmos 561 (1973-33A)	25 May 1341	65	206	295	Third generation; low-resolution cameras; emit PDM^a signals; ejected capsule
Cosmos 563 (1973-36A)	6 Jun 1131	65	206	298	Third generation; high-resolution cameras; manoeuvrable; emit morse code signals; ejected capsule
Cosmos 572 (1973-38A)	10 Jun 1019	52	206	281	Third generation; high-resolution cameras; manoeuvrable; emit morse code signals; ejected capsule
Cosmos 575 (1973-43A)	21 Jun 1326	65	204	271	Third generation; low-resolution cameras; emit PDM^a signals
Cosmos 576 (1973-44A)	27 Jun 1200	73	204	332	Third generation; low-resolution cameras; two-tone telemetry transmission; ejected capsule
Cosmos 577 (1973-48A)	25 Jul 1131	65	207	289	Third generation; high-resolution cameras; manoeuvrable; emit morse code signals; ejected capsule
Cosmos 578 (1973-51A)	1 Aug 1410	65	200	292	Third generation; low-resolution cameras; emit PDM^a signals
Cosmos 579 (1973-55A)	21 Aug 1229	65	196	282	Third generation; high-resolution cameras; manoeuvrable; emit morse code signals; ejected capsule
Cosmos 581 (1973-59A)	24 Aug 1117	52	208	288	Third generation; high-resolution cameras; manoeuvrable; emit morse code signals; ejected capsule
Cosmos 583 (1973-62A)	30 Aug 1033	65	204	298	Third generation; low-resolution cameras; emit PDM^a signals
Cosmos 584 (1973-63A)	6 Sep 1048	73	205	336	Third generation; high-resolution cameras; manoeuvrable; emit morse code signals; ejected capsule
Cosmos 587 (1973-66A)	21 Sep 1312	65	205	300	Third generation; high-resolution cameras; two-tone telemetry transmission; manoeuvrable; ejected capsule
Cosmos 596 (1973-70A)	3 Oct 1258	65	206	287	Unclassified; ejected capsule
Cosmos 597 (1973-71A)	6 Oct 1229	65	206	290	Third generation; high-resolution cameras; two-tone telemetry transmission; manoeuvrable; ejected capsule
Cosmos 598 (1973-72A)	10 Oct 1048	73	208	334	Third generation; high-resolution cameras; manoeuvrable; emit morse code signals; ejected capsule
Cosmos 599 (1973-73A)	15 Oct 0853	65	202	280	Third generation; low-resolution cameras; emit PDM^a signals

Satellite name and designation	Launch date and time GMT	Orbital inclination deg	Perigee height km	Apogee height km	Comments
Cosmos 600 (1973-74A)	16 Oct 1214	73	205	340	Third generation; high-resolution cameras; manoeuvrable; emit morse code signals; ejected capsule
Cosmos 602 (1973-77A)	20 Oct 1019	73	210	335	Third generation; high-resolution cameras; two-tone telemetry transmission; manoeuvrable; ejected capsule
Cosmos 603 (1973-79A)	27 Oct 1117	73	205	357	Third generation; high-resolution cameras; manoeuvrable; emit morse code signals; ejected capsule
Cosmos 607 (1973-87A)	10 Nov 1243	73	204	341	Unclassified; ejected capsule
Cosmos 609 (1973-92A)	21 Nov 1005	70	241	314	Third generation; high-resolution cameras; manoeuvrable; emit morse code signals; ejected capsule
Cosmos 612 (1973-95A)	28 Nov 1146	73	206	346	Third generation; high-resolution cameras; two-tone telemetry transmission; manoeuvrable; ejected capsule
Cosmos 616 (1973-102A)	17 Dec 1200	73	206	332	Third generation; low-resolution cameras; two-tone telemetry transmission; ejected capsule
Cosmos 625 (1973-105A)	21 Dec 1229	73	204	321	Third generation; high-resolution cameras; two-tone telemetry transmission; ejected capsule
1974					
Cosmos 629 (1974-03A)	24 Jan 1507	63	197	289	Third generation; low-resolution cameras; emit PDM[a] signals; ejected capsule
Cosmos 630 (1974-04A)	30 Jan 1102	73	203	346	Third generation; high-resolution cameras; two-tone telemetry transmission; manoeuvrable
Cosmos 632 (1974-06A)	12 Feb 0907	65	176	303	Third generation; high-resolution cameras; manoeuvrable; emit morse code signals; ejected capsule
Cosmos 635 (1974-14A)	14 Mar 1033	73	204	326	Third generation; low-resolution cameras; emit PDM[a] signals; ejected capsule
Cosmos 636 (1974-16A)	20 Mar 0838	65	165	386	Third generation; high-resolution cameras; two-tone telemetry transmission; manoeuvrable
Cosmos 639 (1974-19A)	4 Apr 0838	81	206	226	Third generation; high-resolution cameras; two-tone telemetry transmission; manoeuvrable
Cosmos 640 (1974-21A)	11 Apr 1229	81	201	225	Third generation; low-resolution cameras; emit PDM[a] signals

Satellite name and designation	Launch date and time GMT	Orbital inclination deg	Perigee height km	Apogee height km	Comments
Cosmos 649 (1974-27A)	29 Apr 1326	63	181	299	Third generation; high-resolution cameras; two-tone telemetry transmission; manoeuvrable; ejected capsule
Cosmos 652 (1974-30A)	15 May 0838	52	173	343	Third generation; high-resolution cameras; two-tone telemetry transmission; manoeuvrable; ejected capsule
Cosmos 653 (1974-31A)	15 May 1229	63	192	287	Third generation; low-resolution cameras; emit PDM[a] signals
Cosmos 657 (1974-38A)	30 May 1243	63	177	296	Third generation; high-resolution cameras; two-tone telemetry transmission; manoeuvrable
Cosmos 658 (1974-41A)	6 Jun 0629	65	204	286	Third generation; low-resolution cameras; emit PDM[a] signals
Cosmos 659 (1974-43A)	13 Jun 1229	63	153	329	Third generation; high-resolution cameras; two-tone telemetry transmission; manoeuvrable; ejected capsule
Cosmos 664 (1974-49A)	29 Jun 1258	73	205	341	Third generation; low-resolution cameras; two-tone telemetry transmission; ejected capsule
Cosmos 666 (1974-53A)	12 Jul 1258	63	181	328	Third generation; high-resolution cameras; two-tone telemetry transmission; manoeuvrable; ejected capsule
Cosmos 667 (1974-57A)	25 Jul 0658	65	176	320	Unclassified; ejected capsule
Cosmos 669 (1974-59A)	26 Jul 0658	81	209	230	Third generation; low-resolution cameras; emit PDM[a] signals; ejected capsule
Cosmos 671 (1974-62A)	7 Aug 1258	63	182	353	Third generation; high-resolution cameras; two-tone telemetry transmission; manoeuvrable
Cosmos 674 (1974-68A)	29 Aug 0735	65	175	323	Third generation; high-resolution cameras; two-tone telemetry transmission; manoeuvrable; ejected capsule
Cosmos 685 (1974-73A)	20 Sep 0936	65	205	285	Third generation; low-resolution cameras; emit PDM[a] signals
Cosmos 688 (1974-78A)	18 Oct 1507	63	179	349	Third generation; high-resolution cameras; two-tone telemetry transmission; manoeuvrable; ejected capsule
Cosmos 691 (1974-82A)	25 Oct 0936	65	173	328	Third generation; high-resolution cameras; two-tone telemetry transmission; manoeuvrable; ejected capsule

Satellite name and designation	Launch date and time GMT	Orbital inclination deg	Perigee height km	Apogee height km	Comments
Cosmos 692 (1974-87A)	1 Nov 1424	63	197	295	Third generation; low-resolution cameras; emit PDM[a] signals; ejected capsule
Cosmos 693 (1974-88A)	4 Nov 1048	81	219	243	Third generation; low-resolution cameras; two-tone telemetry transmission; ejected capsule
Cosmos 694 (1974-90A)	16 Nov 1146	73	173	313	Third generation; high-resolution cameras; two-tone telemetry transmission; manoeuvrable
Cosmos 696 (1974-95A)	27 Nov 1146	73	205	321	Third generation; low-resolution cameras; emit PDM[a] signals
Cosmos 697 (1974-98A)	13 Dec 1341	63	174	392	Unclassified
Cosmos 701 (1974-106A)	27 Dec 0907	71	205	319	Third generation; high-resolution cameras; two-tone telemetry transmission; manoeuvrable; ejected capsule
1975					
Cosmos 702 (1975-02A)	17 Jan 0907	71	205	313	Third generation; low-resolution cameras; emit PDM[a] signals
Cosmos 704 (1975-05A)	23 Jan 1102	73	205	305	Third generation; high-resolution cameras; two-tone telemetry transmission; manoeuvrable
Cosmos 709 (1975-13A)	12 Feb 1438	63	181	310	Third generation; high-resolution cameras; two-tone telemetry transmission; manoeuvrable[c]
Cosmos 710 (1975-15A)	26 Feb 0907	65	176	335	Third generation; high-resolution cameras; two-tone telemetry transmission; manoeuvrable; ejected capsule
Cosmos 719 (1975-18A)	12 Mar 0853	65	175	307	Third generation; high-resolution cameras; two-tone telemetry transmission; manoeuvrable
Cosmos 720 (1975-19A)	21 Mar 0658	63	212	273	Third generation; low-resolution cameras; two-tone telemetry transmission; ejected capsule[c]
Cosmos 721 (1975-20A)	26 Mar 0853	81	208	228	Third generation; low-resolution cameras; emit PDM[a] signals; ejected capsule[c]
Cosmos 722 (1975-21A)	27 Mar 0810	71	204	337	Third generation; high-resolution cameras; two-tone telemetry transmission; manoeuvrable
Cosmos 727 (1975-30A)	16 Apr 0810	65	172	334	Third generation; high-resolution cameras; two-tone telemetry transmission; manoeuvrable; ejected capsule[c]

Satellite name and designation	Launch date and time *GMT*	Orbital inclination *deg*	Perigee height *km*	Apogee height *km*	Comments
Cosmos 728 (1975-31A)	18 Apr 1005	73	205	323	Third generation; low-resolution cameras; emit PDM^a signals; ejected capsule
Cosmos 730 (1975-35A)	24 Apr 0810	81	210	234	Third generation; high-resolution cameras; two-tone telemetry transmission; manoeuvrable
Cosmos 731 (1975-41A)	21 May 0658	65	203	296	Third generation; low-resolution cameras; emit PDM^a signals; ejected capsule
Cosmo 740 (1975-46A)	28 May 0735	65	173	327	Third generation; high-resolution cameras; two-tone telemetry transmission; manoeuvrable; ejected capsule^c
Cosmos 741 (1975-47A)	30 May 0643	81	210	231	Third generation; low-resolution cameras; emit PDM^a signals
Cosmos 742 (1975-48A)	3 Jun 1326	63	178	355	Third generation; high-resolution cameras; two-tone telemetry transmission; manoeuvrable
Cosmos 743 (1975-53A)	12 Jun 1229	63	181	331	Third generation; high-resolution cameras; two-tone telemetry transmission; manoeuvrable^c
Cosmos 746 (1975-59A)	25 Jun 1258	63	180	325	Third generation; high-resolution cameras; two-tone telemetry transmission; manoeuvrable; ejected capsule^c
Cosmos 747 (1975-60A)	27 Jun 1258	63	193	291	Third generation; low-resolution cameras; emit PDM^a signals; ejected capsule^c
Cosmos 748 (1975-61A)	3 Jul 1341	63	178	317	Third generation; high-resolution cameras; two-tone telemetry transmission; manoeuvrable
Cosmos 751 (1975-68A)	23 Jul 1258	63	197	313	Third generation; low-resolution cameras; emit PDM^a signals
Cosmos 753 (1975-71A)	31 Jul 1258	63	181	330	Third generation; high-resolution cameras; two-tone telemetry transmission; manoeuvrable^c
Cosmos 754 (1975-73A)	13 Aug 0726	71	204	326	Third generation; high-resolution cameras; two-tone telemetry transmission; manoeuvrable
Cosmos 757 (1975-78A)	27 Aug 1453	63	182	316	Third generation; high-resolution cameras; two-tone telemetry transmission; manoeuvrable^c
Cosmos 758 (1975-80A)	5 Sep 1453	67	174	326	Possible fourth generation with longer orbital lifetime; exploded in orbit
Cosmos 759 (1975-84A)	12 Sep 0531	63	231	276	Third generation; low-resolution cameras; two-tone telemetry transmission; ejected capsule^c

Satellite name and designation	Launch date and time GMT	Orbital inclination deg	Perigee height km	Apogee height km	Comments
Cosmos 760 (1975-85A)	16 Sep 0907	65	174	335	Third generation; high-resolution cameras; two-tone telemetry transmission; manoeuvrable
Cosmos 769 (1975-88A)	23 Sep 1005	73	203	307	Third generation; low-resolution cameras; emit PDM[a] signals; ejected capsule[c]
Cosmos 771 (1975-90A)	25 Sep 0950	81	203	219	Third generation; high-resolution cameras; two-tone telemetry transmission; manoeuvrable
Cosmos 774 (1975-95A)	1 Oct 0838	71	204	315	Third generation; high-resolution cameras; two-tone telemetry transmission; manoeuvrable
Cosmos 776 (1975-101A)	17 Oct 1438	63	200	288	Third generation; low-resolution cameras; emit PDM[a] signals; ejected capsules
Cosmos 779 (1975-104A)	4 Nov 1522	63	182	341	Third generation; high-resolution cameras; two-tone telemetry transmission; manoeuvrable
Cosmos 780 (1975-108A)	21 Nov 0922	65	201	278	Third generation; low-resolution cameras; emit PDM[a] signals; ejected capsule[c]
Cosmos 784 (1975-113A)	3 Dec 1000	81	215	232	Third generation; low-resolution cameras; emit PDM[a] signals; ejected capsule[c]
Cosmos 786 (1975-120A)	16 Dec 1000	65	174	326	Third generation; high-resolution cameras; two-tone telemetry transmission; manoeuvrable[c]
1976					
Cosmos 788 (1976-02A)	7 Jan 1536	63	183	321	Third generation; high-resolution cameras; two-tone telemetry transmission; manoeuvrable
Cosmos 799 (1976-09A)	29 Jan 0838	71	205	306	Third generation; low-resolution cameras; emit PDM[a] signals
Cosmos 802 (1976-13A)	11 Feb 0853	65	172	334	Third generation; high-resolution cameras; two-tone telemetry transmission; manoeuvrable
Cosmos 805 (1976-18A)	20 Feb 1410	67	171	351	Possibly fourth generation with longer orbital lifetime
Cosmos 806 (1976-20A)	10 Mar 0810	71	178	334	Second generation; low-resolution cameras
Cosmos 809 (1976-25A)	18 Mar 0922	65	205	300	Third generation; low-resolution cameras; emit PDM[a] signals
Cosmos 810 (1976-28A)	26 Mar 1507	63	181	338	Third generation; high-resolution cameras; two-tone telemetry transmission; manoeuvrable

Satellite name and designation	Launch date and time *GMT*	Orbital inclination *deg*	Perigee height *km*	Apogee height *km*	Comments
Cosmos 811 (1976-30A)	31 Mar 1258	73	206	338	Third generation; high-resolution cameras; two-tone telemetry transmission; manoeuvrable; ejected capsule
Cosmos 813 (1976-33A)	9 Apr 0838	81	210	236	Third generation; low-resolution cameras; emit PDM[a] signals
Cosmos 815 (1976-36A)	28 Apr 0936	81	218	231	Third generation; high-resolution cameras; two-tone telemetry transmission; manoeuvrable; ejected capsule
Cosmos 817 (1976-40A)	5 May 0755	65	173	324	Third generation; high-resolution cameras; two-tone telemetry transmission; manoeuvrable
Cosmos 819 (1976-45A)	20 May 0658	65	202	293	Third generation; low-resolution cameras; emit PDM[a] signals
Cosmos 820 (1976-46A)	21 May 0658	81	209	217	Third generation; high-resolution cameras; two-tone telemetry transmission; manoeuvrable
Cosmos 821 (1976-48A)	26 May 0907	73	204	314	Third generation; high-resolution cameras; two-tone telemetry transmission; manoeuvrable
Cosmos 824 (1976-52A)	8 Jun 0712	71	204	325	Third generation; high-resolution cameras; two-tone telemetry transmission; manoeuvrable
Cosmos 833 (1976-55A)	16 Jun 1312	63	180	316	Third generation; high-resolution cameras; two-tone telemetry transmission; manoeuvrable; ejected capsule
Cosmos 834 (1976-58A)	24 Jun 0712	81	216	237	Third generation; low-resolution cameras; emit PDM[a] signals
Cosmos 835 (1976-60A)	29 Jun 0726	65	174	317	Third generation; high-resolution cameras; two-tone telemetry transmission; manoeuvrable; ejected capsule
Cosmos 840 (1976-68A)	14 Jul 0907	73	203	319	Third generation; low-resolution cameras; emit PDM[a] signals
Cosmos 844 (1976-72A)	22 Jul 1550	67	172	353	Possibly fourth generation with longer orbital lifetime; exploded in orbit
Cosmos 847 (1976-79A)	4 Aug 1326	63	181	321	Third generation; high-resolution cameras; two-tone telemetry transmission; manoeuvrable
Cosmos 848 (1976-82A)	12 Aug 1341	63	206	303	Third generation; low-resolution cameras; emit PDM[a] signals
Cosmos 852 (1976-86A)	28 Aug 0907	65	173	332	Third generation; high-resolution cameras; two-tone telemetry transmission; manoeuvrable

Satellite name and designation	Launch date and time GMT	Orbital inclination deg	Perigee height km	Apogee height km	Comments
Cosmos 854 (1976-90A)	3 Sep 0922	81	167	308	Third generation; high-resolution cameras; two-tone telemetry transmission; manoeuvrable
Cosmos 855 (1976-95A)	21 Sep 1146	73	202	341	Third generation; high-resolution cameras; two-tone telemetry transmission; manoeuvrable; ejected capsule
Cosmos 856 (1976-96A)	22 Sep 0936	65	203	300	Third generation; low-resolution cameras; emit PDM[a] signals; ejected capsule
Cosmos 857 (1976-97A)	24 Sep 1507	63	179	323	Third generation; high-resolution cameras; two-tone telemetry transmission; manoeuvrable
Cosmos 859 (1976-99A)	10 Oct 0936	65	173	337	Third generation; high-resolution cameras; two-tone telemetry transmission; manoeuvrable
Cosmos 863 (1976-106A)	25 Oct 1438	63	178	348	Third generation; high-resolution cameras; two-tone telemetry transmission; manoeuvrable
Cosmos 865 (1976-109A)	1 Nov 1131	73	203	326	Third generation; low-resolution cameras; emit PDM[a] signals
Cosmos 866 (1976-110A)	11 Nov 1048	65	180	287	Third generation; high-resolution cameras; two-tone telemetry transmission; manoeuvrable
Cosmos 867 (1976-111A)	23 Nov 1634	63	352	401	Third generation; high-resolution cameras; two-tone telemetry transmission; manoeuvrable
Cosmos 879 (1976-119A)	9 Dec 1005	81	213	225	Third generation; low-resolution cameras; emit PDM[a] signals
Cosmos 884 (1976-123A)	17 Dec 0936	65	166	319	Third generation; high-resolution cameras; two-tone telemetry transmission; manoeuvrable

[a] Pulse Duration Modulation (PDM).

[b] This satellite is included because its orbital characteristics were similar to those of photographic reconnaissance satellites and because its telemetry was in most respects identical to that of satellites which are normally recovered after 12 days but which do not manoeuvre.

[c] During the last orbit, signals were received by the Kettering Group, but these were not the recovery beacon signals.

Table 3A.3. Possible Chinese photographic reconnaissance satellites

Satellite name and designation	Launch date and time *GMT*	Orbital inclination *deg*	Perigee height *km*	Apogee height *km*	Comments
	1975				
China 4 (1975-111A)	26 Nov 0336	63	179	479	Data capsule recovered
	1976				
China 7 (1976-117A)	7 Dec 0430	50	172	489	Possibly only part of payload recovered

Appendix 3B

Tables of electronic reconnaissance satellites

For abbreviations, acronyms and conventions, see page xv.

The designation of each satellite is recognized internationally and is given by the World Warning Agency on behalf of the Committee on Space Research.

More detailed tables of US and Soviet electronic reconnaissance satellites are to be found in *World Armaments and Disarmament, SIPRI Yearbooks 1973* (pp. 84–86, 89, 99–101), *1974* (pp. 299, 301), *1975* (pp. 398, 400), *1976* (p. 118) and *1977* (pp. 138, 156).

Table 3B.1. US electronic reconnaissance satellites

Satellite name and designation	Launch date and time *GMT*	Orbital inclination *deg*	Perigee height *km*	Apogee height *km*	Comments
1962					
USAF (1962-δ1)	21 Feb ..	82	167	374	
USAF (1962-ω1)	18 Jun 2024	82	370	441	
Star-rad (1962-βπ1)	26 Oct 1619	71	194	5 537	
USAF/USN (1962-βτ1)	13 Dec 0405	70	231	2 786	
1963					
USAF (1963-03A)	16 Jan 2248	82	459	533	
USAF —	26 Apr —	—	—	—	Failed to orbit
USAF (1963-21E)	15 Jun 1438	70	181	829	
USAF (1963-27A)	29 Jun 2234	82	484	536	
USAF (1963-35B)	29 Aug 1912	82	310	431	
USAF (1963-41B)	25 Oct 1858	99	136	297	
USAF (1963-42B)	29 Oct 2107	90	285	585	First-known octagonal satellite in group
USAF (1963-55B)	21 Dec 2150	65	321	388	

Satellite name and designation	Launch date and time GMT	Orbital inclination deg	Perigee height km	Apogee height km	Comments
1964					
USAF (1964-01A)	11 Jan 2010	70	905	934	
USAF (1964-01E)	11 Jan 2010	70	905	934	
USAF (1964-11A)	28 Feb 0338	82	479	520	
USAF (1964-35A)	3 Jul 0126	82	501	529	
USAF (1964-36B)	6 Jul 2150	93	297	377	
USAF (P II) (1964-45B)	14 Aug 2248	96	275	3 748	
USAF (1964-68B)	23 Oct 1829	96	323	336	
USAF (1964-72A)	4 Nov 0134	82	512	526	
1965					
SR 6B (1965-16A)	9 Mar 1829	70	910	939	First 8-payload launch
USAF (1965-31B)	28 Apr 2010	95	490	559	
USAF (1965-50A)	24 Jun 1926	108	496	510	
USAF (1965-55A)	17 Jul 0600	70	471	512	
USAF (1965-62B)	3 Aug 1912	107	501	515	
USAF	2 Sep	—	—	—	Failed to orbit
1966					
USAF (1966-02A)	19 Jan 2010	94	150	269	
USAF (1966-09A)	9 Feb 2010	82	508	512	
USAF (1966-22A)	18 Mar 2024	101	162	308	
USAF (1966-39B)	19 Apr 1912	110	517	559	
USAF (1966-48B)	3 Jun 1926	87	136	281	
USAF (1966-74B)	16 Aug 1929	93	510	524	
USAF (1966-83B)	16 Sep 1800	94	460	501	
USAF (1966-90B)	12 Oct 1912	91	181	258	

Satellite name and designation	Launch date and time GMT	Orbital inclination deg	Perigee height km	Apogee height km	Comments
USAF (1966-98B)	2 Nov 2024	91	208	324	
USAF (1966-118A)	29 Dec 1200	75	486	496	
1967					
USAF (1967-43B)	9 May 2150	85	555	809	
USAF (1967-50B)	22 May 1829	92	148	240	
USAF (1967-53E)	31 May 0936	70	916	921	
USAF (1967-53G)	31 May 0926	70	915	927	
USAF (1967-53H)	31 May 0936	70	915	926	
USAF (1967-62B)	16 Jun 2136	80	501	517	
USAF (1967-71A)	25 Jul 0350	75	458	513	
USAF (1967-109B)	2 Nov 2136	82	455	524	
1968					
USAF (1968-04A)	17 Jan 1005	75	450	546	
USAF (1968-08B)	24 Jan 2219	82	473	542	
USAF (1968-20B)	14 Mar 2245	83	481	522	
USAF (1968-52B)	20 Jun 2150	85	437	519	
USAF (1968-78A)	18 Sep 2136	83	500	514	
USAF (1968-86A)	5 Oct 1117	75	483	511	
USAF (1968-112B)	12 Dec 2219	80	1 391	1 468	
1969					
USAF (1969-10B)	5 Feb 2248	80	1 396	1 441	
USAF (1969-26B)	19 Mar 2136	83	504	513	
USAF (1969-41B)	2 May 0155	66	401	473	
USAF (1969-65A)	31 Jul 1005	75	462	541	

Satellite name and designation	Launch date and time GMT	Orbital inclination deg	Perigee height km	Apogee height km	Comments
USAF (1969-79B)	22 Sep 2107	85	490	496	
USAF (1969-82A)	30 Sep 1341	70	446	484	
1970					
USAF (1970-16B)	4 Mar 2219	88	442	514	
USAF (1970-40B)	20 May 2136	83	491	503	
USAF (1970-66A)	26 Aug 1005	75	484	504	
USAF (1970-98B)	18 Nov 2122	83	487	511	
1971					
USAF (1971-60A)	16 Jul 1048	75	488	508	
USAF (1971-76B)	10 Sep 2136	75	492	507	
USAF (1971-110A)	14 Dec 1214	70	983	999	Several other possible electronic reconnaissance satellites launched at same time
1972					
USAF (1972-02D)	20 Jan 1836	97	472	549	First satellite launched from Big Bird
USAF (1972-52C)	7 Jul 1746	96	497	504	
USAF (1972-79C)	10 Oct 1800	96	1 423	1 469	
1973					
USAF (1973-88B)	10 Nov 2010	96	486	508	
USAF (1973-88D)	10 Nov 2010	97	1 419	1 458	
1974					
USAF (1974-20B)	10 Apr 0824	95	786	830	
USAF (1974-20C)	10 Apr 0824	94	503	531	
USAF (1974-85B)	29 Oct 1926	96	520	535	
1975					
USAF (SSU-1) (1975-51C)	8 Jun 1829	95	1 389	1 401	

Satellite name and designation	Launch date and time *GMT*	Orbital inclination *deg*	Perigee height *km*	Apogee height *km*	Comments
USAF (1975-114B)	4 Dec 2038	96	236	1 558	
	1976				
USAF (1976-65C)	8 Jul 1843	96	628	632	

Table 3B.2. Possible Soviet electronic reconnaissance satellites

Satellite name and designation	Launch date and time *GMT*	Orbital inclination *deg*	Perigee height *km*	Apogee height *km*	Comments
	1967				
Cosmos 148 (1967-23A)	16 Mar 1746	71	270	404	
Cosmos 152 (1967-28A)	25 Mar 0658	71	272	488	
Cosmos 173 (1967-81A)	24 Aug 0502	71	277	480	
Cosmos 189 (1967-108A)	30 Oct 1800	74	524	565	
Cosmos 191 (1967-115A)	21 Nov 1424	71	267	497	
	1968				
Cosmos 200 (1968-06A)	19 Jan 2205	74	523	537	
Cosmos 204 (1968-15A)	5 Mar 1117	71	275	844	
Cosmos 222 (1968-44A)	30 May 2024	71	285	488	
Cosmos 242 (1968-79A)	20 Sep 1438	71	272	406	
Cosmos 245 (1968-83A)	3 Oct 1258	71	284	473	
Cosmos 250 (1968-95A)	31 Oct 2205	74	522	542	
Cosmos 257 (1968-107A)	3 Dec 1453	71	286	462	
	1969				
Cosmos 265 (1969-12A)	7 Feb 1410	71	275	458	
Cosmos 269 (1969-21A)	5 Mar 1731	74	525	543	

Satellite name and designation	Launch date and time *GMT*	Orbital inclination *deg*	Perigee height *km*	Apogee height *km*	Comments
Cosmos 275 (1969-31A)	28 Mar 1605	71	273	780	
Cosmos 277 (1969-33A)	4 Apr 1258	71	268	466	
Cosmos 285 (1969-49A)	3 Jun 1258	71	267	493	
Cosmos 295 (1969-73A)	22 Aug 1424	71	270	469	
Cosmos 303 (1969-90A)	18 Oct 1005	71	270	466	
Cosmos 308 (1969-96A)	4 Nov 1200	71	271	408	
Cosmos 311 (1969-102A)	24 Nov 1102	71	273	469	
Cosmos 314 (1969-106A)	11 Dec 1258	71	272	465	
Cosmos 315 (1969-107A)	20 Dec 0336	74	518	542	
1970					
Cosmos 324 (1970-14A)	27 Feb 1731	71	275	466	
Cosmos 327 (1970-20A)	19 Mar 1438	71	280	819	
Cosmos 330 (1970-24A)	7 Apr 1117	74	514	543	
Cosmos 334 (1970-33A)	23 Apr 1326	71	272	482	
Cosmos 351 (1970-51A)	27 Jun 0741	71	270	467	
Cosmos 357 (1970-63A)	19 Aug 1507	71	272	476	
Cosmos 362 (1970-73A)	16 Sep 1200	71	270	829	
Cosmos 369 (1970-81A)	8 Oct 1507	71	269	506	
Cosmos 387 (1970-111A)	16 Dec 0434	74	528	538	
Cosmos 388 (1970-112A)	18 Dec 0936	71	271	505	
1971					
Cosmos 391 (1971-02A)	14 Jan 1200	71	267	803	
Cosmos 393 (1971-07A)	26 Jan 1243	71	272	485	
Cosmos 395 (1971-13A)	18 Feb 2107	74	529	546	

Satellite name and designation	Launch date and time GMT	Orbital inclination deg	Perigee height km	Apogee height km	Comments
Cosmos 421 (1971-44A)	19 May 1019	71	273	469	
Cosmos 423 (1971-47A)	27 May 1200	71	272	487	
Cosmos 425 (1971-50A)	29 May 0350	74	506	553	
Cosmos 435 (1971-72A)	27 Aug 1102	71	271	482	
Cosmos 436 (1971-74A)	7 Sep 0126	74	509	545	
Cosmos 437 (1971-75A)	10 Sep 0350	74	519	548	
Cosmos 440 (1971-79A)	24 Sep 1034	71	272	785	
Cosmos 453 (1971-90A)	19 Oct 1243	71	271	493	
Cosmos 455 (1971-97A)	17 Nov 1117	71	272	491	
Cosmos 458 (1971-101A)	29 Nov 1019	71	272	497	
Cosmos 460 (1971-103A)	30 Nov 1648	74	528	532	
Cosmos 467 (1971-113A)	17 Dec 1048	71	267	472	
1972					
Cosmos 479 (1972-17A)	22 Mar 2035	74	514	543	
Cosmos 498 (1972-50A)	5 Jul 0936	71	267	490	
Cosmos 500 (1972-53A)	10 Jul 1619	74	505	549	
Cosmos 523 (1972-78A)	5 Oct 1131	71	272	481	
Cosmos 524 (1972-80A)	11 Oct 1326	71	267	512	
Cosmos 526 (1972-84A)	25 Oct 1048	71	273	486	
Cosmos 536 (1972-88A)	3 Nov 0141	74	518	544	
1973					
Cosmos 544 (1973-03A)	20 Jan 0336	74	510	548	
Cosmos 545 (1973-04A)	24 Jan 1146	71	269	495	Satellite decayed
Cosmos 549 (1973-10A)	28 Feb 0434	74	513	545	

Satellite name and designation	Launch date and time GMT	Orbital inclination deg	Perigee height km	Apogee height km	Comments
Cosmos 553 (1973-20A)	12 Apr 1200	71	272	494	Satellite decayed
Cosmos 558 (1973-29A)	17 May 1326	71	269	501	
Cosmos 562 (1973-35A)	5 Jun 1131	71	270	487	
Cosmos 580 (1973-57A)	22 Aug 1131	71	273	493	
Cosmos 582 (1973-60A)	28 Aug 1005	74	519	543	
Cosmos 608 (1973-91A)	20 Nov 1229	71	270	503	
Cosmos 610 (1973-93A)	27 Nov 0014	74	515	546	
Cosmos 611 (1973-94A)	28 Nov 1005	71	270	481	
Cosmos 615 (1973-99A)	13 Dec 1117	71	270	834	
1974					
Cosmos 631 (1974-05A)	6 Feb 0043	74	521	545	
Cosmos 633 (1974-10A)	27 Feb 1117	71	271	491	
Cosmos 634 (1974-12A)	5 Mar 1605	71	271	491	
Cosmos 655 (1974-35A)	21 May 0614	74	523	542	
Cosmos 661 (1974-45A)	21 Jun 0907	74	511	548	
Cosmos 662 (1974-47A)	26 Jun 1229	71	271	812	
Cosmos 668 (1974-58A)	25 Jul 1200	71	270	492	
Cosmos 686 (1974-74A)	26 Sep 1634	71	273	489	
Cosmos 695 (1974-91A)	20 Nov 1200	71	273	468	
Cosmos 698 (1974-100A)	18 Dec 1410	74	515	552	
1975					
Cosmos 705 (1975-06A)	28 Jan 1200	71	271	502	
Cosmos 707 (1975-08A)	5 Feb 1326	74	503	547	
Cosmos 725 (1975-26A)	8 Apr 1829	71	270	481	

Satellite name and designation	Launch date and time GMT	Orbital inclination deg	Perigee height km	Apogee height km	Comments
Cosmos 745 (1975-58A)	24 Jun 1214	72	264	514	
Cosmos 749 (1975-62A)	4 Jul 0058	74	509	556	
Cosmos 750 (1975-67A)	17 Jul 0907	71	272	803	
Cosmos 781 (1975-109A)	21 Nov 1717	74	505	551	
1976					
Cosmos 787 (1976-01A)	6 Jan 0502	74	518	547	
Cosmos 790 (1976-07A)	22 Jan 2234	74	511	549	
Cosmos 801 (1976-12A)	5 Feb 1435	71	268	796	
Cosmos 812 (1976-31A)	6 Apr 0419	74	508	548	
Cosmos 818 (1976-44A)	18 May 1102	71	271	481	
Cosmos 845 (1976-75A)	27 Jul 0531	74	514	546	
Cosmos 849 (1976-83A)	18 Aug 0936	71	264	865	
Cosmos 850 (1976-84A)	26 Aug 1102	71	272	493	
Cosmos 870 (1976-115A)	2 Dec 0014	74	513	548	

Appendix 3C

Tables of ocean-surveillance satellites

For abbreviations, acronyms and conventions, see page xv.

The designation of each satellite is recognized internationally and is given by the World Warning Agency on behalf of the Committee on Space Research.

More detailed tables of US and Soviet ocean-surveillance satellites are to be found in *World Armaments and Disarmament, SIPRI Yearbooks 1976* (p. 118) and *1977* (pp. 138,157).

Table 3C.1. US ocean-surveillance satellites

Satellite name and designation	Launch date and time *GMT*	Orbital inclination *deg*	Perigee height *km*	Apogee height *km*	Comments
1976					
USN NOSSI (1976-38A)	30 Apr 1912	64	1 092	1 128	
USN SSU-1 (1976-38C)	30 Apr 1912	63	1 093	1 129	
USN SSU-2 (1976-38D)	30 Apr 1912	63	1 093	1 130	
USN SSU-3 (1976-38J)	30 Apr 1912	64	1 083	1 139	

Table 3C.2. Possible Soviet ocean-surveillance satellites[a]

Satellite name and designation	Launch date and time *GMT*	Orbital inclination *deg*	Perigee height *km*	Apogee height *km*	Comments
1967					
Cosmos 198 (1967-127A)	27 Dec 1131	65 65	249 894	270 952	
1968					
Cosmos 209 (1968-23A)	22 Mar 0936	65 65	183 871	343 944	
1970					
Cosmos 367 (1970-79A)	3 Oct 1033	65 65	250 922	280 1 024	See reference [35a]

91

Satellite name and designation	Launch date and time *GMT*	Orbital inclination *deg*	Perigee height *km*	Apogee height *km*	Comments
1971					
Cosmos 402	1 Apr	65	247	274	
(1971-25A)	1131	65	948	1 036	
Cosmos 469	25 Dec	65	249	262	
(1971-117A)	1131	65	941	1 023	
1972					
Cosmos 516	21 Aug	65	251	263	
(1972-66A)	1033	65	920	1 030	
1973					
Cosmos 626	27 Dec	65	257	259	
(1973-108A)	2024	65	910	990	
1974					
Cosmos 651	15 May	65	250	264	
(1974-29A)	0726	65	892	954	
Cosmos 654	17 May	65	248	265	
(1974-32A)	0658	65	913	1 024	
1975					
Cosmos 723	2 Apr	65	249	266	
(1975-24A)	1102	65	916	951	
Cosmos 724	7 Apr	65	248	266	
(1975-25A)	1102	65	870	934	
Cosmos 785	12 Dec	65	251	261	
(1975-116A)	1300	65	898	1 023	
1976					
Cosmos 860	17 Oct	65	252	265	
(1976-103A)	1814	65	919	1 008	
Cosmos 861	21 Oct	65	251	265	
(1976-104A)	1702	65	919	1 005	

[a] The second figure under orbital inclination, perigee height and apogee height is that of the final orbit.

Appendix 3D

Tables of early-warning satellites

For abbreviations, acronyms and conventions, see page xv.

The designation of each satellite is recognized internationally and is given by the World Warning Agency on behalf of the Committee on Space Research.

More detailed tables of US and Soviet early-warning satellites are to be found in *World Armaments and Disarmaments, SIPRI Yearbooks 1973* (pp. 86–89), *1974* (p. 299), *1976* (p. 115) and *1977* (pp. 138, 157).

Table 3D.1. US MIDAS, Vela and other early-warning satellites

Satellite name and designation	Launch date and time *GMT*	Orbital inclination *deg*	Perigee height *km*	Apogee height *km*	Comments
MIDAS					
	1960				
MIDAS 1 —	26 Feb —	—	—	—	Failed to orbit
MIDAS 2 (1960-ξ1)	24 May 1731	33	484	511	
Discoverer 19 (1960-τ)	20 Dec 2238	83	209	631	
	1961				
Discoverer 21 (1961-ζ)	18 Feb 2248	81	240	1069	
MIDAS 3 (1961-σ1)	12 Jul 1619	91	3358	3534	
MIDAS 4 (1961-αδ1)	12 Oct 1961	96	3496	3756	
	1962				
MIDAS 5 (1962-χ1)	9 Apr 1550	87	2814	3382	
USAF —	17 Dec —	—	—	—	Failed to orbit
	1963				
MIDAS 6 (1963-14A)	9 May 2010	87	3604	3680	
USAF —	12 Jun —	—	—	—	Failed to orbit

Satellite name and designation	Launch date and time GMT	Orbital inclination deg	Perigee height km	Apogee height km	Comments
MIDAS 7 (1963-30A)	19 Jul 0350	88	3 670	3 727	
1966					
USAF (1966-77A)	19 Aug 1926	90	3 680	3 700	
USAF (1966-89A)	5 Oct 2248	90	3 682	3 702	

Vela

1963					
Vela 1 (1963-39A)	17 Oct 0224	38	102 098	111 137	
Vela 2 (1963-39C)	17 Oct 0224	38	99 300	115 800	
1964					
Vela 3 (1964-40A)	17 Jul 0824	40	101 959	104 591	
Vela 4 (1964-40B)	17 Jul 0824	41	94 436	111 775	
1965					
Vela 5 (1965-58A)	20 Jul 0824	35	88 534	96 238	
Vela 6 (1965-58B)	20 Jul 0824	35	101 859	121 453	
1967					
Vela 7 (1967-40A)	28 Apr 1005	30	107 337	114 612	
Vela 8 (1967-40B)	28 Apr 1005	30	107 337	114 612	
1969					
Vela 9 (1969-46D)	23 May 0755	33	110 900	112 210	
Vela 10 (1969-46E)	23 May 0755	33	110 920	112 283	
1970					
Vela 11 (1970-27A)	8 Apr 1104	32	111 210	112 160	
Vela 12 (1970-27B)	8 Apr 1104	33	111 500	112 210	

Satellite name and designation	Launch date and time *GMT*	Orbital inclination *deg*	Perigee height *km*	Apogee height *km*	Comments
Other early-warning					
	1966				
USAF (1966-51A)	9 Jun 2010	90	174	3 616	
	1968				
USAF (1968-63A)	6 Aug 1117	10	31 680	39 860	
	1969				
BMEWS 2 (1969-36A)	13 Apr 0224	10	32 670	39 270	
	1970				
BMEWS 3 (1970-46A)	19 Jun 1131	28	178	33 685	
BMEWS 4 (1970-69A)	1 Sep 0058	10	31 680	39 860	
IMEWS 1 (1970-93A)	6 Nov 1083	8	26 050	35 886	
	1971				
IMEWS 2 (1971-39A)	5 May 0755	1	35 651	35 840	
	1972				
IMEWS 3 (1972-10A)	1 Mar 0938	1	35 678	35 871	
USAF (1972-101A)	20 Dec	98	14 000	14 000	
	1973				
BMEWS 6 (1973-13A)	6 Mar 1200	10	42 259	32 100	
IMEWS 4 (1973-40A)	12 Jun 0936	1	35 533	35 901	
	1975				
BMEWS ? (1975-55A)	18 Jun 1005	9	32 200	40 800	
IMEWS 5 (1975-118A)	14 Dec 0517	1	35 620	35 860	
	1976				
USAF (1976-59A)	26 Jun 0307	1	35 620	35 860	

Table 3D.2. Possible Soviet early-warning satellites

Satellite name and designation	Launch date and time *GMT*	Orbital inclination *deg*	Perigee height *km*	Apogee height *km*	Comments
1967					
Cosmos 159 (1967-46A)	16 May 2150	52	350	60 637	Possible precursor test of manned spacecraft
Cosmos 174 (1967-82A)	31 Aug 0755	65	430	39 796	Possible Molniya failure
1968					
Cosmos 260 (1968-115A)	16 Dec 0922	65	518	39 570	
1972					
Cosmos 520 (1972-72A)	19 Sep 1926	63	750	39 470	
1973					
Cosmos 606 (1973-84A)	2 Nov 1258	63	657	39 310	
1974					
Cosmos 665 (1974-50A)	29 Jun 1605	63	625	39 378	
1975					
Cosmos 706 (1975-07A)	30 Jan 1507	63	623	39 824	
Cosmos 775 (1975-97A)	8 Oct 0029	0	35 737	36 220	
1976					
Cosmos 862 (1976-105A)	22 Oct 0922	63	571	39 516	

4. Communications satellites

The highly sophisticated and powerful modern weapons of today's world have increased the need of the military for fast and efficient communications. For such weapons to be credible and for them to be secure from unauthorized use, reliable and effective command and control links are essential.

Conventional communications systems use both undersea and short-wave radio, both of which are limited in efficiency. Undersea cables provide service only to those parts of the world which are linked by the cable network. In radio communications systems, high-frequency signals are often degraded by violent electrical disturbances in the atmosphere and ionosphere.

The USA and the USSR have for the past decade and a half used satellite communications systems which are capable of maintaining reliable and effective command and control, probably involving the use of more than one satellite in each system.

Communication by satellite has considerable advantages from the military point of view in that it enables the military to communicate simultaneously with terminals which are thousands of miles apart as well as with those which may be just over the next hill. This flexibility is further enhanced by the availability of transportable or even mobile ground terminals. With such a system, the satellite transponder (receiver–amplifier–transmitter) replaces several intermediate repeaters needed for long-distance communications in the Earthbound conventional microwave system. However, one problem with a satellite communications system is that it is visible over a wide area so that it is available for use by others and ordinary transmissions may be easily interfered with or jammed.

Communications satellites can be classified into two broad categories, according to their electromagnetic and orbital characteristics. The first category comprises both passive and active satellites. A passive satellite, which may be a large metallic skin balloon construction, acts only as a reflector of radio waves. The Moon, for example, has been used as a passive communications satellite. One serious disadvantage with such a system is that the reflected signal is very weak by the time it reaches the terminal on Earth. Thus, passive satellites are not commonly used and are therefore now only of historical interest. An active satellite, on the other hand, carries a transponder system which receives communications signals transmitted from ground stations and amplifies them and retransmits them to other Earth stations.

In the second category, communications satellites can be classified according to their orbital characteristics into three general types: synchronous, semi-synchronous and non-synchronous satellites.

I. Satellite orbits

Most of the global communications satellite systems which have been planned or used in the United States make use of synchronous orbits. The satellite orbital inclination in this case is, by definition, zero; that is, the orbital plane coincides with the equatorial plane (see Figure 2.3, page 6). The orbital period of the satellite is 24 h and the satellite's altitude is 35 900 km. On the other hand, a non-synchronous satellite reaches an altitude other than 35 900 km and orbits round the Earth with a period which depends on its altitude (Figure 4.1). The non-synchronous satellite is, therefore, visible from a given point on the Earth during only a part of its orbit; the duration and frequency of the visibility will depend on the orbital characteristics and on the positions of the Earth terminals. From some terminals there may be good visibility during only some orbits, while from other terminals the satellites may not be seen at all. Therefore, many non-synchronous satellites are needed for continuous coverage of the Earth.

Figure 4.1. Satellite periods for various altitudes

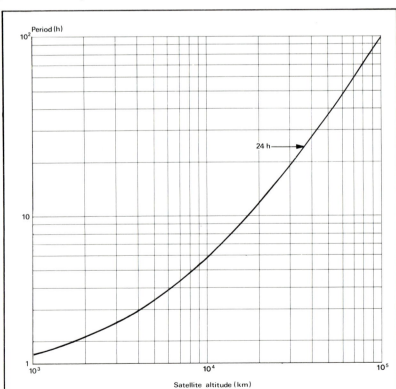

The total area of the Earth covered by a satellite at any time is bounded by a circle the radius of which is a function of satellite altitude h and minimum allowable Earth antenna-elevation angle α (Figure 4.2). In the case of an equatorial synchronous satellite, the Earth coverage area is fixed, whereas for a non-synchronous satellite in a circular orbit the coverage area is circular, fixed in size and moving continuously over the Earth's surface.

The coverage or spherical area A_s of the Earth's surface within the visibility cone of angle 2θ (Figure 4.2) is given by

$$A_s = 2\pi R_e^2 (1 - \cos \theta) \tag{4.1}$$

Figure 4.2. Satellite coverage geometry. $\varphi = 90°(\alpha + \theta)$; $h =$ height of satellite above the Earth; $R_s =$ slant range, terminal-to-satellite; $R_e =$ radius of Earth; $\alpha =$ elevation angle at terminal; $\theta =$ angular radius of visibility; C = centre of the Earth; $A_s =$ coverage area.

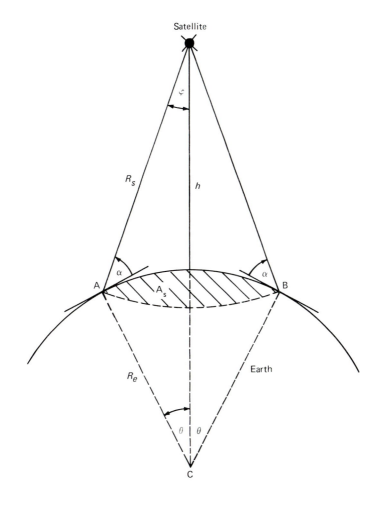

Figure 4.3. Changes of percentage of Earth's surface visible from different satellite altitudes

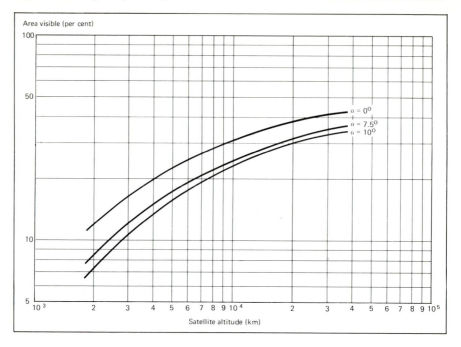

where

$$\theta = \left[\cos^{-1} \left| \frac{R_e \cos \alpha}{R_e + h} \right| \right] - \alpha.$$

The percentage of the Earth's surface in view from a satellite at various altitudes is shown in Figure 4.3 for several values of α. When $\alpha = 0°$, the spherical area becomes

$$A_s = 2\pi R_e^2 [h/(R_e + h)].$$

It is important that at an Earth terminal the duration of visibility of a satellite is long. This duration increases as the satellite's altitude increases and also as the satellite orbital plane moves closer to the Earth terminal. At any altitude a direct overhead pass results in the longest duration of visibility. If a communications satellite has no information-storage capability, the satellite must be able to see two Earth terminals simultaneously to establish a link. This makes a non-synchronous satellite system more complex than a synchronous system.

Some of the problems encountered by communications satellites in various types of orbits are illustrated by considering some examples. On 12 August 1960, Echo 1 was launched by the United States in a circular orbit with an altitude of about 1600 km. The orbit of this passive communications satellite was unstable. Solar radiation pressure acting on the satellite during the few

100

months of its lifetime made its orbit less circular, but in June 1961 the orbit became circular again. This pattern was repeated several times and each time the satellite travelled into denser atmosphere at its perigee, thus increasing the air drag on the satellite and shortening its lifetime. Such satellites are therefore unsuitable for long-distance communications since they are very sensitive to changes in air drag and radiation pressure and are difficult to track accurately.

Active communications satellites such as Telstar also experience the radiation pressure of direct sunlight as well as that reflected from the Earth. The radiation pressure of sunlight is constant and can be compensated for, but that reflected from the Earth as well as the infra-red radiation emitted by the Earth vary unpredictably. However, the effects of these on a heavy satellite are small and therefore cause no substantial changes in the satellite's orbit.

Military communications satellites in the United States are placed in circular orbits at heights of about 35 000 km so that the effects of atmospheric drag are minimized. These satellites have an orbital inclination of nearly $0°$ so that their orbital period is about 24 h and they therefore follow the Earth's rotation, hovering over one spot above the Earth. An important advantage of such a synchronous orbit for long-distance communications satellites is that they can be tracked by an almost stationary aerial rather than by a rapidly moving one. Since the signals have to travel at least 72 000 km, however, sensitive radio receivers must be used. On the other hand, a satellite in such an orbit can be seen from about one-third of the Earth's surface.

One disadvantage of such an orbit for countries situated far from the equator is that, for latitudes north of 70°N, the satellite is below the horizon and therefore unusable. The USSR therefore chose for its Molniya communications satellites a semi-synchronous orbit with a 12-h period and perigee and apogee heights of about 500 km and 40 000 km, respectively. The apogee is over the northern hemisphere so that the satellite spends most of its time in orbit in high northern latitudes. One effect of the Earth's flattening at the poles and bulging at the equator is to cause the satellite orbit to turn in its own plane. However, if the orbital inclination is 63.4°, then the orbit remains stationary and will have a stable perigee. The orbital inclinations of most Soviet Molniya satellites are close to 63.4°, so that the movement of the apogee is negligible throughout the active lifetime of the satellite.

Since these satellites travel at high altitudes (about 40 000 km) they are considerably affected by the gravitational forces of the Sun and Moon. The satellites have to make small manoeuvres to stay in position. Such corrections also have to be made to synchronous satellites since there is a sudden change in the gravitational field of the Earth in the mid-Atlantic region. With such orbital manoeuvres, satellites remain stationary over this region. Ideal positions for synchronous satellites are over the Pacific and Indian Oceans.

The selection of a particular orbit must also be subject to any limitations on the sensors the satellite carries. For a communications satellite, an important limitation of radio systems is the restriction on range imposed by transmission power limits. Satellite weight is nearly proportional to power so that more satellites can be launched with a given vehicle if the power requirements are low or if the satellite can be placed in a high-altitude orbit. Moreover, low power tends to reduce interference. For a particular rate of signal transmission, almost the same power will serve for an oriented satellite with a directional antenna at an altitude of 35 900 km as that required for an unoriented satellite at an altitude of 3200–8000 km radiating equally in all directions.

Below an altitude of about 1600 km, an active communications satellite provides adequate service with omnidirectional antennas and relatively simple subsystems. For intercontinental communications, the lowest useful altitude is about 3200 km. But at such altitudes and with circular orbits, hundreds of satellites would be required for significant coverage. The number of satellites necessary can be reduced if they are orbited above about 8000 km. For high-quality communications, active-repeater satellites (see below) need to be used, for which a minimum altitude of 1600 km is necessary [57]. The best altitude is one at which the smallest number of satellites is required for worldwide coverage.

For global coverage, the cost of construction of the ground network is inversely proportional to the satellite altitude. More stations are required for point-to-point worldwide communications for satellites at low altitudes. A minimum of ground stations is required for satellites in 24 h synchronous orbits.

II. Satellite transponder characteristics

A satellite transponder receives signals, amplifies them and retransmits them to other Earth stations. If a repeater is operated on the same frequency band for both input and output signals, these signals can be isolated by using different frequency bands for each type of signal. The frequency conversion is usually achieved by using either single or double conversion: the choice depends on the required input-to-output power gain and the channel bandwidth [57]. If the bandwidth is large, a single conversion may be adequate. Where the input and output operating frequencies are high and narrow bandwidth is required, a double-conversion transponder may be needed.

The transponder is sometimes designed to include signal processing. The input radio-frequency (RF) signal is demodulated and the baseband signal is then modulated onto the output RF carrier. For military satellites, such a system can improve protection against jamming [58].

In addition to the transponder, a communications satellite also has a system for transmitting a beacon signal to allow the Earth terminal antennas to locate and track the satellite. This signal may be generated within the satellite. Sometimes a beacon signal is produced by converting the frequency of a signal transmitted by the Earth terminal. Often, however, the beacon and telemetry functions are combined by modulating the telemetry information onto the beacon signal. The transponder in the US Defense Satellite Communications System (DSCS) Phase I satellites is a double-frequency conversion repeater, transmitting at 7.3 GHz and receiving at 8.0 GHz [58].

In order to receive and transmit signals, the transponder requires an antenna. The choice of antenna depends on its frequency response, bandwidth, polarization, gain and beamwidth. A basic characteristic of an antenna, whether used for transmitting or receiving signals, is its gain. The gain G of an antenna relative to an isotropic antenna is given by

$$G = A/A_{iso} = A4\pi/\lambda^2 \tag{4.2}$$

where

λ = wavelength of signal,
A = effective area of antenna = true area of antenna $\times \eta$,
η = efficiency of the antenna,

and

$A_{iso}(\lambda^2/4\pi)$ = effective area of an isotropic antenna.

The most commonly used directive antenna in ultra-high-frequency and microwave point-to-point applications is the parabolic antenna. The efficiency η of such an antenna is less than unity, usually about 0.55.

Therefore, the gain of a parabolic antenna is

$$G = \eta(\pi d/\lambda)^2 \tag{4.3}$$

or

$$G_{dB} = 10 \log [\eta(\pi d/\lambda)^2]$$

where

G_{dB} = aperture gain in dB

and

d = antenna diameter in the same units as λ.

Another important limitation of radio systems for communications is the restriction on range due to transmission power limits. For simple one-way transmission, the range equation takes several forms:

$$\begin{aligned} P_r/P_t &= G_t A_r/4\pi R^3 \\ &= G_r A_t/4\pi R^2 \\ &= G_t G_r \lambda^2/(4\pi R)^2 \\ &= A_t A_r/(\lambda R)^2 \end{aligned} \tag{4.4}$$

where

P_r = received signal power,
P_t = transmitted power,
A_r = effective area of the receiving antenna,
A_t = effective area of the transmitting antenna,
G_r = receiving antenna gain,
G_t = transmitting antenna gain,

and

R = range from the transmitter to the receiver.

In the Echo passive satellite, a portion of the transmitted radio beam is reflected by the satellite. A fraction of the reflected energy is then received at the transmitter station. If A_o is the scattering cross-section area, then

$$P_r/P_t = A_o A_r G_t/(4\pi R^2)^2$$

$$= G_t G_r \lambda^2 A_o/[4\pi(4\pi R^2)^2]. \tag{4.5}$$

Amplifying a signal does not ensure that it will be detected because the level of noise in a system may obscure the signal; that is, theoretically, if no interference were present, any signal transmitted over any distance could be detected provided there were sufficient amplification in the receiver. The source of the noise limiting the usable range may be at the transmitter, at the receiver or in the space link (propagating medium). The equivalent noise power is usually expressed as a noise temperature T in degrees kelvin.

The noise power produced in a bandwidth Δf is given by

$$P_n = kT\Delta f \tag{4.6}$$

where

k = Boltzmann's constant,
T = absolute temperature of the circuitry in kelvin,

and

Δf = bandwidth (hertz).

The temperature T is the weighted sum of the various component source temperatures, the weights being high (unity) for noise sources internal to or surrounding the receiver or low if the noise enters through a fraction of the receiver surroundings. Equivalent temperatures of external noise are dependent on frequency and in the case of atmospheric noise, on elevation angle. The black-body radiation of the Earth is also a source of noise which increases for low-elevation angles. Therefore, low-altitude satellites which spend a relatively long time near the horizon entail greater noise problems for an observer on Earth.

A perfect receiver, for example, would have $T = 0$ K, a fair one $T = 293$ K, and a poor one $T = 2930$ K. Typical equivalent temperatures of internal receiver noise are 2000 K for conventional receivers, 100 K for parametric amplifier receivers and 10 K for masers [5].

The ratio of receiver power P_r to interference power in the bandwidth Δf determines the range at which communication is feasible. A ratio of unity is defined as the threshold reception. The threshold reception range R for space-to-Earth communication is

$$R = \sqrt{(P_t A_t G_r / 4\pi k T \Delta f)}. \tag{4.7}$$

The usable range may be about one-third of this value.

III. The US programme

The US DoD has basically two independent types of communications satellite systems forming part of the World-Wide Military Command and Control System (WWMCCS). These are often called strategic and tactical communications satellite systems. The first system carries high data-rate strategic command and control signals over long distances, using large ground and shipboard terminals. Moreover, intelligence data, high-priority warning and special communications are also transmitted by such systems. In the tactical system, satellites are employed to transmit essential command and control communications necessary for military forces. This type of system can only process low data rate. It has been reported that plans exist to use satellites for communication between stations in the United States and individuals gathering intelligence in other parts of the world [59].

The USAF and USN are the agencies mainly concerned with the development of communications satellite systems. The USAF programme will be considered first.

The concept of the communications satellite system began in 1958 with Project SCORE (Signal Communication by Orbiting Relay Equipment) when the USAF orbited an Atlas ICBM-type burnt-out rocket stage equipped with two-way radio equipment which transmitted taped messages for 12 days. The SCORE satellite was followed by Echo passive satellites in 1960 and in 1964. This latter type of satellite requires ground-station transmitters of much higher power than those required for active satellites. An active satellite, Courier, was also launched in 1960.

After a long series of studies, the Initial Defense Communication Satellite Program (IDCSP) was established towards the end of 1964 by the USAF [60]. Under this programme two types of system were suggested. In one it was planned to use Atlas–Agena boosters to place a number of satellites at a time into approximately 11 000-km random polar orbits. In a second system, only a few synchronous-altitude satellites would be placed in equatorial orbits. In the latter case, only two satellites would be launched at a time using Titan-3C

Figure 4.4. Artist's impression of a US DSCS II communications satellite

rocket. After a successful Titan-3C launch in June 1965, however, the Atlas–Agena project was cancelled and in June 1966, seven IDCSP satellites were launched in near-synchronous orbits using a single Titan-3C rocket. The IDCSP was renamed the Initial Defense Satellite Communication System (IDSCS). While the IDSCS provided limited operational capability between widely separated fixed terminals, the LES (Lincoln Experimental Satellite) satellite series demonstrated advances in technology which allowed communication between aircraft, ships, mobile ground terminals and large fixed installations [60].

The second phase of the DSCS system is a high-capacity, super-high frequency (SHF) system which provides protection from jammed voice and data links for the WWMCCS. This system also provides for the worldwide Diplomatic Telecommunication Service as well as providing transmission of some surveillance, intelligence and early-warning data. Under this programme the USAF has launched two satellites in 1971, two in late 1973 and two in May 1975. Figure 4.4 shows one such satellite.

It has been argued that a reliable and survivable satellite communications system for command, control and communications of US strategic nuclear forces must be able to withstand severe physical and jamming attacks during wartime if the strategic nuclear force is to be a credible deterrent [61]. Considerable effort has been spent on this question under the Air Force Satellite Communication (AFSATCOM) programme; two experimental satellites, LES 8 and 9, were launched in 1975 to demonstrate and validate the necessary technology. The AFSATCOM will use short, low-speed (teletype) messages for force execution, for report-back and force redirection. The use of such messages together with suitable anti-jamming techniques will allow for relatively simple UHF low-power terminals aboard operational vehicles.

The first regularly operating satellite telecommunication service, a service which still continues, was demonstrated by the USN in January 1960. As further advances in the field of UHF and higher frequency propagation and their applications to communications satellites were made, the USN expanded its programme to use man-made satellites for communications. With this technique, transmission of signals became relatively insensitive to atmospheric and solar disturbances. In 1970, the USN evaluated its satellite communications programme by testing Tacsat 1 (Tactical Satellite) and LES 6 [62]. The latter satellite is still in orbit providing limited operation use.

In 1975, the USN leased a satellite from the Communications Satellite Corporation (Comsat) [62] until various other satellites became operational. It is planned that, beginning in late 1976 and over a period of one year, the USN will launch three or four geostationary satellites. These satellites will be called the Fleet Satellite Communication (FLTSATCOM) and will provide communications by digitalized voice, teleprinter and other techniques. Moreover, these satellites will be able to transmit computer-to-computer data thus providing real-time readiness, for example, of ocean surveillance and combat direction. The FLTSATCOM will operate at UHF but, unlike the DSCS, it will have a relatively small capacity. These satellites are expected to be launched some time in 1977. Orbital data of these satellites are shown in Table 4A.1.

IV. The Soviet programme

It is not possible precisely to determine how extensively the Soviet armed forces use civilian satellites for military purposes. The civilian and military Soviet communications satellite programmes are carried out under the Molniya series (Figure 4.5). With the increasing number of Molniya satellites, it is very likely that the Soviet military use these satellites for their purposes, particularly since domestic television coverage has not been expanded through use of the extra channels available. By the end of 1975, for example, the Molniya 1 series (begun in 1965) consisted of 33 satellites; the Molniya 2 series (from 1971) consisted of 15 satellites; and the Molniya 3 series (from 1974) consisted of three satellites.

Figure 4.5. The Soviet communications satellite Molniya 1

Unlike US satellites, Soviet spacecraft carry large payloads of at least 1000 kg and have about 10 times the power output of the US Early Bird satellites of the same period. In the Soviet space programme the basic launch vehicle is their 1957 ICBM, the SS-6 Sapwood, to which are added one or more upper stages, depending on the mission. For the Molniya flights, the vehicle consists of a $1\frac{1}{2}$-stage booster with a second-generation upper stage plus escape stage; the vehicle has been designated the A-2-e.

The Soviet Union places its communications satellites in orbits with periods of about 12 h, with perigee heights of about 500 km in the southern hemisphere and apogee heights of about 40 000 km in the northern hemisphere. If three such satellites are orbited 120° apart in a plane, each satellite will provide about nine hours of coverage per day over the Soviet Union. Because of the orbital inclination of 62.8°, the satellite not only provides excellent coverage at northern latitudes but also provides visibility simultaneously on passes through the apogee across the polar regions. The satellites with orbital planes 120° apart were superseded by those with orbital planes spaced 90° apart. This constituted a four-satellite communications system. During 1976 the Molniya 1 satellites have been placed in between Molniya 2 and 3. This leads to speculation that Molniya 1 satellites are military communications satellites [63].

Molniya satellites make two orbits daily, one of which is over the Soviet Union and the other over North America. The orbital parameters are optimized so that the longest communication period occurs over the region between Moscow and Vladivostok. The first Molniya satellite was a cylinder which carried six solar battery panels and had two parabolic aerials mounted on it

(Figure 4.5). The antennas are folded during launch and automatically open out after the carrier rocket has separated out. Radio-electronic equipment is carried inside the cylinder. During the entire flight the satellite is oriented with its solar batteries facing the Sun and under operational conditions one of the aerials is directed towards the Earth. The second aerial is kept in reserve. The satellite can handle a television broadcast, a large number of telephone conversations, still pictures and telegraph communications and can relay other forms of information.

It has been the intention of the Soviet Union for some time to use 24-h synchronous satellites, which have been used regularly by the United States. The first Soviet synchronous satellite, Cosmos 637, was orbited in 1974. Later that year, Molniya 1S was placed in an equatorial synchronous orbit and it was not until the end of 1975 that a new series of such synchronous satellites began with the launch of Statsionar 1 (Raduga).

Although long-distance communications are facilitated by a satellite in an equatorial synchronous orbit that can be tracked by an almost stationary aerial rather than by a rapidly guided one, a drawback for the Soviet Union is that the greater proportion of its land mass lies far from the equator and since the satellite is below the horizon for latitudes north of 70°N it is therefore unusable. It is thus not surprising that Molniya satellites use 12-h orbits.

The formation of the Soviet communications network using synchronous orbits probably began with the launch of Statsionar 1. It is possible that the Soviet Union is planning to launch at least an 11-satellite network into synchronous orbit by about 1980 since it informed the Frequency Registration Board of the International Telecommunications Union that it plans to launch the Statsionar-T satellite for domestic television communications, and Statsionar 2 and 3 for overall Soviet and European communications. It has been announced that Statsionar 4-10 may be launched in 1978–80 [64]. The orbital location of Statsionar-T will be 99°E longitude with the Earth-to-satellite link within 6.2 GHz $\pm$ 12 MHz and satellite-to-Earth link within the range 714 MHz $\pm$ 12 MHz. Statsionar 2 is to be located at 35°E longitude over the eastern part of Africa for communications services to Europe and the Western part of the Soviet Union. The Earth-to-satellite link will be in the frequency range 5.75–6.2 GHz and that from satellite-to-Earth will be 3.42–3.87 GHz. The system is designed for telephone, telegraph and phototelegraph communications and for sound and television broadcasting. Statsionar 3 is similar to Statsionar 2 except that it will serve the whole of the Soviet Union (apart from the extreme north and Kamchatka). It will be positioned at 85°E longitude over the southern part of India [65]. Statsionar 4-10 satellites are planned to operate on 4-GHz and 6-GHz frequency bands used by Intelsat satellites. Statsionar 4 will be placed at 14°W longitude, Statsionar 5 at 58°E longitude and Statsionar 10 at 170°W longitude. Statsionar 8 and 9, at 25°W longitude and 45°E longitude, respectively, will reinforce northern hemispheric coverage while Statsionar 6 and 7 at 85°E longitude and 140°E longitude, respectively, will primarily cover the domestic telecommunications services [66].

The orbital characteristics of these satellites are given in Table 4A.2.

V. The British programme

On 19 September 1966, as a result of a request from the British government, the USA and the UK agreed that the USAF would procure a synchronous communications satellite for the UK. It was also agreed that the USAF would launch the satellite into the required orbit and then turn the command and control over to the UK. On 22 November 1969, Skynet 1 was launched by a Delta rocket into a synchronous orbit. A standby was launched on 19 August 1970 but the satellite failed to achieve the required orbit. These two satellites were to have provided the UK with military satellite communications for three to five years and the satellites were to have been replaced by two other satellites at the end of 1973. However, Skynet 2 failed in January 1973 and Skynet 1 ceased to function in January 1972.

It was not until 23 November 1974 that Skynet 2B (Figure 4.6) was successfully launched into a synchronous orbit from Cape Kennedy (ETR). This was

Figure 4.6. The British communications satellite Skynet 2B, shown mated to the three-stage Delta vehicle

the first military communications satellite mainly to be built by a British company. Skynet 2A, launched on 17 January 1974, failed to enter synchronous orbit because the second stage of the Delta launch vehicle failed. Skynet 2B is stationed over the Seychelles in the Indian Ocean and will provide communications in an area bounded by Norway, the Antarctic, western Australia and the Atlantic out to about 23°W.

Communications are carried out over two channel bandwidths, one at 20 MHz and the other at 2 MHz. When a signal is received by the satellite, it is converted to an intermediate frequency of 70 MHz and divided into two channels. Signals in each channel are separately amplified and limited and then recombined. A beacon signal fed into the system is used for tracking the satellite. The finally combined communication and beacon signal is transmitted back to Earth. Such a double conversion provides protection against interference.

The orbital characteristics of these satellites are given in Table 4A.3.

VI. The NATO programme

On 20 March 1970, using a Thor/Delta launcher, NATO put its first communications satellite into a near equatorial synchronous orbit. The satellite was launched by the United States from ETR. It was positioned over the eastern Atlantic, linking the capital cities of the NATO countries. The USAF was responsible for producing and launching the satellite as well as for initially controlling it in orbit.

In late 1966, the United States offered other NATO countries the opportunity of exploring the potential of satellites for tactical military communications. This resulted in a meeting of representatives of seven NATO countries, held in June 1967 at the US Army Satellite Communications Agency, to consider the extent of participation. Criteria for NATO participation in the US programme were formulated at later meetings in Bonn. In November 1967, an understanding was reached among the seven countries (Belgium, Canada, the Federal Republic of Germany, Italy, the Netherlands, the UK and the USA) officially sponsoring a cooperative programme for tactical satellite communications (Tacsatcom). The satellite was to be built and launched by the United States and special ground terminals were to be built by the participants. The programme resulted in the launching of LES 5 (on July 1967) and LES 6 (on September 1968 [67].

The second back-up satellite, NATO 2, was launched from ETR on 3 February 1971, again using a Thor/Delta launcher. The initial testing of the satellite's communications system was carried out by the USAF Space and Missiles Systems Organization (SAMSO). More detailed testing was done by the Signals Research and Development Establishment in the UK and then by the SHAPE Technical Centre in the Hague [68].

Figure 4.7. The NATO communications satellite NATO 3B, being readied for launch

NATO 2 covers an area from the eastern coast of North America to the eastern boundary of Turkey. The communications system is basically designed to operate with only one satellite, the other being a standby satellite. The more recent NATO satellite to be launched is the second of three NATO 3 satellites (NATO 3B): NATO 3A, launched from ETR on 22 April 1976, was the largest communications satellite developed for NATO. NATO 3B (Figure 4.7) was launched on 28 January 1977 from ETR.

The orbital characteristics of these satellites are given in Table 4A.4.

VII. The French programme

It has been reported that in 1977 the armed forces technical services will carry out military space communications technology experiments using the French–

Figure 4.8. The French–West German communications satellite Symphonie

West German communications satellite Symphonie (Figure 4.8). Two Symphonie satellites (see Table 4A.5) have been launched by NASA using Thor/Delta rockets. These experiments are planned in collaboration with the Service Central des Télécommunications et de l'Informatique, the Direction Technique des Constructions et Armes Navales and the Direction Technique des Engins (DTEN) [69]. This project is called 'Sextius'.

The orbital characteristics of these satellites are given in Table 4A.5.

Appendix 4A

Tables of communications satellites

For abbreviations, acronyms and conventions, see page xv.

The designation of each satellite is recognized internationally and is given by the World Warning Agency on behalf of the Committee on Space Research.

More detailed tables of communications satellites are to be found in *World Armaments and Disarmament, SIPRI Yearbook 1977* (pp. 141–48, 159–68, 175–76).

Table 4A.1. US communications satellites

Satellite name and designation	Launch date and time *GMT*	Orbital inclination *deg*	Perigee height *km*	Apogee height *km*	Comments
	1958				
ARPA Score (1958-ξ)	18 Dec 2324	32	185	1 484	
	1960				
NASA Echo A-10 —	13 May ..	—	—	—	Failed to orbit
NASA Echo 1 (1960-21)	12 Aug 0936	47	1 524	1 684	
ARPA Courier 1A —	18 Aug ..	—	—	—	Failed to orbit
USA Courier 1B (1960-ν1)	4 Oct 1746	28	938	1 237	
	1961				
USN Lofti (1961-η)	22 Feb 0350	28	167	1 002	
	1962				
NASA Telstar 1 (1962-αε1)	10 Jul 0838	45	952	5 632	
NASA Relay 1 (1962-βν2)	13 Dec 2331	48	1 345	7 398	
	1963				
NASA Syncom 1 (1963-04A)	14 Feb 0517	33	34 392	36 739	

114

Satellite name and designation	Launch date and time *GMT*	Orbital inclination *deg*	Perigee height *km*	Apogee height *km*	Comments
NASA Telstar 2 (1963-13A)	7 May 1131	43	974	10 803	
USAF/USN Lofti 2A (1963-21B)	15 Jun 1438	70	171	925	
NASA Syncom 2 (1963-31A)	26 Jul 1428	33	35 584	36 693	
1964					
NASA Relay 2 (1964-03A)	21 Jan 2107	46	2 091	7 411	
NASA Echo 2 (1964-04A)	25 Jan 1355	82	1 029	1 316	
NASA Syncom 3 (1964-47A)	19 Aug 1214	0	34 191	36 271	
1965					
USAF LES 1 (1965-08C)	11 Feb 1717	32	2 774	2 811	
CSC/NASA (Intelsat 1A) Early Bird (1965-28A)	6 Apr 2346	0	35 003	36 606	
USAF LES 2 (1965-34B)	6 May 1214	32	2 784	14 798	
USAF LES 4 (1965-108B)	21 Dec 1536	27	189	33 632	
USAF LES 3 (1965-108D)	21 Dec 1536	27	195	33 177	
1966					
USAF IDCSP 1 (1965-53B)	16 Jun 1355	0	33 656	33 897	
USAF IDCSP 2 (1966-53C)	16 Jun 1353	0	33 668	33 909	
USAF IDCSP 3 (1966-53D)	16 Jun 1353	0	33 695	33 936	
USAF IDCSP 4 (1966-53E)	16 Jun 1353	0	33 696	34 018	
USAF IDCSP 5 (1966-53F)	16 Jun 1353	0	33 699	34 102	
USAF IDCSP 6 (1966-53G)	16 Jun 1353	0	33 722	34 206	
USAF IDCSP 7 (1966-53H)	16 Jun 1353	0	33 712	34 539	
USAF IDCSP (8 satellites) —	26 Aug ..	—	—	—	Failed to orbit

Satellite name and designation	Launch date and time GMT	Orbital inclination deg	Perigee height km	Apogee height km	Comments
CSC/NASA Intelsat 2A (1966-96A)	26 Oct 2324	26	289	37 656	
USAF OV4 1R (1966-99B)	3 Nov 1355	33	291	298	
USAF OV4 1T (1966-99D)	3 Nov 1355	33	294	321	
NASA ATS 1 (1966-110A)	7 Dec 0210	0	35 852	36 887	
1967					
CSC/NASA Intelsat 2B (1967-01A)	11 Jan 1048	2	35 563	36 496	
USAF IDCSP 8 (1967-03A)	18 Jan 1424	0	33 557	33 800	
USAF IDCSP 9 (1967-03B)	18 Jan 1424	0	33 526	33 846	
USAF IDCSP 10 (1967-03C)	18 Jan 1424	0	33 579	33 819	
USAF IDCSP 11 (1967-03D)	18 Jan 1424	0	33 606	33 847	
USAF IDCSP 12 (1967-03E)	18 Jan 1424	0	33 608	33 929	
USAF IDCSP 13 (1967-03F)	18 Jan 1424	0	33 656	33 978	
USAF IDCSP 14 (1967-03G)	18 Jan 1424	0	33 675	34 077	
USAF IDCSP 15 (1967-03H)	18 Jan 1424	0	33 665	34 229	
CSC/NASA Intelsat 2C (1967-26A)	23 Mar 0126	1	35 687	35 771	
NASA ATS 2 (1967-31A)	6 Apr 0322	28	178	11 124	
USAF IDCSP 16 (1967-66A)	1 Jul 1312	7	32 906	33 528	
USAF IDCSP 17 (1967-66B)	1 Jul 1312	7	33 006	33 548	
USAF IDCSP 18 (1967-66C)	1 Jul 1312	7	33 079	33 555	
USAF DATS 1 (1967-66D)	1 Jul 1312	7	33 156	33 553	
USAF LES 5 (1967-66E)	1 Jul 1312	7	33 178	33 636	
CSC/NASA Intelsat 2D (1967-94A)	25 Sep 0043	1	35 747	35 913	

Satellite name and designation	Launch date and time GMT	Orbital inclination deg	Perigee height km	Apogee height km	Comments
NASA ATS 3 (1967-111A)	5 Nov 2331	1	35 791	36 130	
1968					
USAF IDCSP 19 (1968-50A)	13 Jun 1410	0	33 758	33 841	
USAF IDCSP 20 (1968-50B)	13 Jun 1410	0	33 725	33 863	
USAF IDCSP 21 (1968-50C)	13 Jun 1410	0	33 699	33 907	
USAF IDCSP 22 (1968-50D)	13 Jun 1410	0	33 737	33 954	
USAF IDCSP 23 (1968-50E)	13 Jun 1410	0	33 721	34 035	
USAF IDCSP 24 (1968-50F)	13 Jun 1410	0	33 724	34 126	
USAF IDCSP 25 (1968-50G)	13 Jun 1410	0	33 721	34 256	
USAF IDCSP 26 (1968-50H)	13 Jun 1410	0	33 752	34 443	
NASA ATS 4 (1968-68A)	10 Aug 2234	29	219	726	
CSC/NASA Intelsat 3A —	18 Sep ..	—	—	—	Failed to orbit
USAF LES 6 (1968-81D)	26 Sep 0735	3	35 597	35 785	
CSC/NASA Intelsat 3B (1968-116A)	19 Dec 0029	1	35 770	35 790	
1969					
CSC/NASA Intelsat 3C (1969-11A)	6 Feb 0043	1	35 782	35 808	
USAF Tacsat 1 (1969-13A)	9 Feb 2107	1	35 768	35 803	
CSC/NASA Intelsat 3D (1969-45A)	22 May 0155	29	396	36 093	
CSC/NASA Intelsat 3E (1969-64A)	26 Jul 0210	30	271	5 397	
NASA ATS 5 (1969-69A)	12 Aug 1102	3	35 760	36 894	

Satellite name and designation	Launch date and time GMT	Orbital inclination deg	Perigee height km	Apogee height km	Comments
1970					
CSC/NASA Intelsat 3F (1970-03A)	15 Jan 0014	1	35 773	35 801	
CSC/NASA Intelsat 3G (1970-32A)	23 Apr 0043	0	35 772	35 805	
CSC/NASA Intelsat 3H (1970-55A)	23 Jul 2317	1	19 400	36 030	
1971					
CSC/NASA Intelsat 4A (1971-06A)	26 Jan 0043	1	35 779	35 794	
USAF (1971-21A)	21 Mar 0350	63	390	33 800	Payload may later have injected itself into an inclined synchronous orbit; first launch of this type
USAF DSCS 1 (1971-95A)	3 Nov 0307	3	35 065	36 475	
USAF DSCS 2 (1971-95B)	3 Nov 0307	2	35 349	36 299	
CSC/NASA Intelsat 4B (1971-116A)	20 Dec 0112	0	35 749	35 828	
1972					
CSC/NASA Intelsat 4C (1972-03A)	23 Jan 0014	0	35 781	35 794	
CSC/NASA Intelsat 4D (1972-41A)	13 Jun 2150	0	35 782	35 794	
1973					
USAF (1973-56A)	21 Aug 1605	63	460	39 296	First such launch was in 1971; orbit similar to Soviet communications satellites
CSC/NASA Intelsat 4E (1973-58A)	23 Aug 2324	0	35 539	35 927	
USAF DSCS 3 (1973-100A)	14 Dec 0000	3	35 790	35 791	
USAF DSCS 4 (1973-100B)	14 Dec 0000	3	35 797	35 801	

118

Satellite name and designation	Launch date and time *GMT*	Orbital inclination *deg*	Perigee height *km*	Apogee height *km*	Comments
1974					
WS/NASA Westar 1 (1974-22A)	13 Apr 2331	0	35 761	35 770	
NASA ATS 6 (1974-39A)	30 May 1258	2	35 781	35 791	
WS/NASA Westar 2 (1974-75A)	10 Oct 2248	0	35 710	35 734	
CSC/NASA Intelsat 4F (1974-93A)	21 Nov 2346	2	35 775	35 801	
1975					
CSC/NASA Intelsat 4F-6 —	20 Feb ..	—	—	—	Failed to orbit
USAF SDS 1 (1975-17A)	10 Mar 0448	64	295	39 337	
USAF DSCS 5-6 (1975-40A)	20 May 1410	29	150	249	Failed to reach equatorial synchronous orbit
CSC/NASA Intelsat 4G (1975-42A)	22 May 2248	0	35 780	35 795	
CSC/NASA Intelsat 4A(F-1) (1975-91A)	26 Sep 0014	0	35 780	35 795	
RAC/NASA Satcom 1 (1975-117A)	13 Dec 0155	0	35 625	36 086	
1976					
CSC/NASA Intelsat 4A(F-2) (1976-10A)	30 Jan 0000	0	35 752	35 819	
NASA Marisat 1 (1976-17A)	19 Feb 2234	2	35 703	35 867	
USAF LES 8 (1976-23A)	15 Mar 0126	25	35 787	35 787	
USAF LES 9 (1976-23B)	15 Mar 0126	25	35 787	35 787	
RCA/NASA Satcom 2 (1976-29A)	26 Mar 2234	0	35 785	35 789	
NASA Comstar 1A (1976-42A)	13 May 2234	1	35 780	35 794	
USAF SDS 2 ? (1976-50A)	2 Jun ..	63 ?	380 ?	39 315 ?	
NASA Marisat 2 (1976-53A)	10 Jun 0014	3	35 788	35 807	
NASA Comstar 1B (1976-73A)	22 Jul 2248	1	35 780	35 795	

Satellite name and designation	Launch date and time GMT	Orbital inclination deg	Perigee height km	Apogee height km	Comments
USAF SDS 3 ? (1976-80A)	6 Aug ..	63·3	380	39 315	
NASA Marisat 3 (1976-101A)	14 Oct 2248	3	35 051	36 525	

Table 4A.2. Possible Soviet communications satellites

Satellite name and designation	Launch date and time GMT	Orbital inclination deg	Perigee height km	Apogee height km	Comments
1964					
Cosmos 41 (1964-49D)	22 Aug 0712	65	426	39 771	Precursor to Molniya 1
Cosmos 42 (1964-50A)	22 Aug 1102	49	224	1 098	Dual launch
Cosmos 43 (1964-50C)	22 Aug 1102	49	227	1 100	Dual launch
1965					
Molniya 1-1 (1965-30A)	23 Apr 0155	66	538	39 300	
Cosmos 80 (1965-70A)	3 Sep 1355	56	1 357	1 555	Quintuple launch
Cosmos 81 (1965-70B)	3 Sep 1355	56	1 384	1 557	Quintuple launch
Cosmos 82 (1965-70C)	3 Sep 1355	56	1 408	1 565	Quintuple launch
Cosmos 83 (1965-70D)	3 Sep 1355	56	1 441	1 567	Quintuple launch
Cosmos 84 (1965-70E)	3 Sep 1355	56	1 466	1 576	Quintuple launch
Molniya 1-2 (1965-80A)	14 Oct 1938	65	481	39 935	
Cosmos 103 (1965-112A)	28 Dec 1229	56	594	636	See reference [25a]
1966					
Molniya 1-3 (1966-35A)	25 Apr 0712	65	506	39 492	
Molniya 1-4 (1966-92A)	20 Oct 0755	65	505	39 685	
1967					
Cosmos 151 (1967-27A)	24 Mar 1146	56	596	652	See reference [25a]

120

Satellite name and designation	Launch date and time GMT	Orbital inclination deg	Perigee height km	Apogee height km	Comments
Cosmos 158 (1967-45A)	15 May 1102	74	738	822	See reference [25a]
Molniya 1-5 (1967-52A)	24 May 2248	65	1 188	38 807	
Molniya 1-6 (1967-95A)	3 Oct 0502	65	502	39 868	
Molniya 1-7 (1967-101A)	22 Oct 0838	65	508	39 710	
1968					
Molniya 1-8 (1968-35A)	21 Apr 0419	65	391	39 738	
Molniya 1-9 (1968-57A)	5 Jul 1522	65	401	39 803	
Cosmos 236 (1968-70A)	27 Aug 1131	56	588	630	See reference [25a]
Molniya 1-10 (1968-85A)	5 Oct 0029	65	436	39 633	
1969					
Molniya 1-11 (1969-35A)	11 Apr 0238	65	404	39 741	
Molniya 1-12 (1969-61A)	22 Jul 1258	65	499	39 519	
1970					
Molniya 1-13 (1970-13A)	19 Feb 1858	65	461	39 170	
Cosmos 336 (1970-36A)	25 Apr 1702	74	1 464	1 490	Octuple launch
Cosmos 337 (1970-36B)	25 Apr 1702	74	1 470	1 554	Octuple launch
Cosmos 338 (1970-36C)	25 Apr 1702	74	1 472	1 518	Octuple launch
Cosmos 339 (1970-36D)	25 Apr 1702	74	1 446	1 472	Octuple launch
Cosmos 340 (1970-36E)	25 Apr 1702	74	1 409	1 473	Octuple launch
Cosmos 341 (1970-36F)	25 Apr 1702	74	1 345	1 471	Octuple launch
Cosmos 342 (1970-36G)	25 Apr 1702	74	1 313	1 471	Octuple launch
Cosmos 343 (1970-36H)	25 Apr 1702	72	1 374	1 474	Octuple launch
Molniya 1-14 (1970-49A)	26 Jun 0322	65	448	39 260	
Molniya 1-15 (1970-77A)	29 Sep 0824	66	480	39 300	

Satellite name and designation	Launch date and time *GMT*	Orbital inclination *deg*	Perigee height *km*	Apogee height *km*	Comments
Cosmos 372 (1970-86A)	16 Oct 1507	74	785	806	See reference [25a]
Molniya 1-16 (1970-101A)	27 Nov 1550	66	471	39 350	
Molniya 1-17 (1970-114A)	25 Dec 0350	65	495	39 565	
1971					
Cosmos 407 (1971-35A)	23 Apr 1131	74	791	819	See reference [25a]
Cosmos 411 (1971-41A)	7 May 1424	74	1 318	1 492	Octuple launch
Cosmos 412 (1971-41B)	7 May 1424	74	1 482	1 537	Octuple launch
Cosmos 413 (1971-41C)	7 May 1424	74	1 476	1 509	Octuple launch
Cosmos 414 (1971-41D)	7 May 1424	74	1 428	1 496	Octuple launch
Cosmos 415 (1971-41E)	7 May 1424	74	1 452	1 503	Octuple launch
Cosmos 416 (1971-41F)	7 May 1424	74	1 373	1 494	Octuple launch
Cosmos 417 (1971-41G)	7 May 1424	74	1 344	1 495	Octuple launch
Cosmos 418 (1971-41H)	7 May 1424	74	1 401	1 495	Octuple launch
Molniya 1-18 (1971-64A)	28 Jul 0336	65	468	39 254	
Cosmos 444 (1971-86A)	13 Oct 1341	74	1 324	1 509	Octuple launch
Cosmos 445 (1971-86B)	13 Oct 1341	74	1 353	1 513	Octuple launch
Cosmos 446 (1971-86C)	13 Oct 1341	74	1 384	1 513	Octuple launch
Cosmos 447 (1971-86D)	13 Oct 1341	74	1 414	1 515	Octuple launch
Cosmos 448 (1971-86E)	13 Oct 1314	74	1 441	1 522	Octuple launch
Cosmos 449 (1971-86F)	13 Oct 1341	74	1 484	1 544	Octuple launch
Cosmos 450 (1971-86G)	13 Oct 1241	74	1 465	1 530	Octuple launch
Cosmos 451 (1971-86H)	13 Oct 1314	74	1 492	1 574	Octuple launch
Molniya 2-1 (1971-100A)	24 Nov 0936	66	517	39 554	
Cosmos 468 (1971-114A)	17 Dec 1258	74	786	809	See reference [25a]

Satellite name and designation	Launch date and time *GMT*	Orbital inclination *deg*	Perigee height *km*	Apogee height *km*	Comments
Molniya 1-19 (1971-115A)	19 Dec 2324	65	499	39 139	
	1972				
Molniya 1-20 (1972-25A)	4 Apr 2038	66	480	39 260	
Molniya 2-2 (1972-37A)	19 May 1438	65	440	39 290	
Cosmos 494 (1972-43A)	23 Jun 0922	74	790	804	See reference [25a]
Cosmos 504 (1972-57A)	20 Jun 1800	74	1 324	1 498	Octuple launch
Cosmos 505 (1972-57B)	20 Jun 1800	74	1 354	1 498	Octuple launch
Cosmos 506 (1972-57C)	20 Jun 1800	74	1 384	1 498	Octuple launch
Cosmos 507 (1972-57D)	20 Jun 1800	74	1 414	1 498	Octuple launch
Cosmos 508 (1972-57E)	20 Jun 1800	74	1 446	1 497	Octuple launch
Cosmos 509 (1972-57F)	20 Jun 1800	74	1 475	1 501	Octuple launch
Cosmos 510 (1972-57G)	20 Jun 1800	74	1 497	1 512	Octuple launch
Cosmos 511 (1972-57H)	20 Jun 1800	74	1 496	1 548	Octuple launch
Molniya 2-3 (1972-57A)	30 Sep 2024	66	392	39 240	
Molniya 1-21 (1972-81A)	14 Oct 0614	65	480	39 300	
Cosmos 528 (1972-87A)	1 Nov 0155	74	1 368	1 471	Octuple launch
Cosmos 529 (1972-87B)	1 Nov 0155	74	1 404	1 470	Octuple launch
Cosmos 530 (1972-87C)	1 Nov 0155	74	1 336	1 469	Octuple launch
Cosmos 531 (1972-87D)	1 Nov 0155	74	1 423	1 471	Octuple launch
Cosmos 532 (1972-87E)	1 Nov 0155	74	1 302	1 470	Octuple launch
Cosmos 533 (1972-87F)	1 Nov 0155	74	1 319	1 470	Octuple launch
Cosmos 534 (1972-87G)	1 Nov 0155	74	1 351	1 470	Octuple launch
Cosmos 535 (1972-87H)	1 Nov 0155	74	1 385	1 472	Octuple launch
Molniya 1-22 (1972-95A)	2 Dec 0448	65	555	39 797	

Satellite name and designation	Launch date and time *GMT*	Orbital inclination *deg*	Perigee height *km*	Apogee height *km*	Comments
Molniya 2-4 (1972-98A)	12 Dec 0658	65	495	39 300	
Cosmos 540 (1972-104A)	25 Dec 2324	74	781	810	See reference [25a]
1973					
Molniya 1-23 (1973-07A)	3 Feb 0600	65	470	39 164	
Molniya 2-5 (1973-18A)	5 Apr 1117	66	477	39 107	
Cosmos 564 (1973-37A)	8 Jun 1536	74	1 395	1 484	Octuple launch
Cosmos 565 (1973-37B)	8 Jun 1536	74	1 450	1 492	Octuple launch
Cosmos 566 (1973-37C)	8 Jun 1536	74	1 435	1 485	Octuple launch
Cosmos 567 (1973-37D)	8 Jun 1536	74	1 414	1 486	Octuple launch
Cosmos 568 (1973-37E)	8 Jun 1536	74	1 378	1 482	Octuple launch
Cosmos 569 (1973-37F)	8 Jun 1536	74	1 359	1 482	Octuple launch
Cosmos 570 (1973-37G)	8 Jun 1536	74	1 341	1 481	Octuple launch
Cosmos 571 (1973-37H)	8 Jun 1536	74	1 321	1 481	Octuple launch
Molniya 2-6 (1973-45A)	11 Jun 1445	65	441	39 285	
Molniya 1-24 (1973-61A)	30 Aug 0014	66	463	39 893	
Cosmos 588 (1973-69A)	2 Oct 2150	74	1 451	1 494	Octuple launch
Cosmos 589 (1973-69B)	2 Oct 2150	74	1 419	1 487	Octuple launch
Cosmos 590 (1973-69C)	2 Oct 2150	74	1 438	1 486	Octuple launch
Cosmos 591 (1973-69D)	2 Oct 2150	74	1 349	1 488	Octuple launch
Cosmos 592 (1973-69E)	2 Oct 2150	74	1 333	1 486	Octuple launch
Cosmos 593 (1973-69F)	2 Oct 2150	74	1 366	1 487	Octuple launch
Cosmos 594 (1973-69G)	2 Oct 2150	74	1 382	1 488	Octuple launch
Cosmos 595 (1973-69H)	2 Oct 2150	74	1 402	1 486	Octuple launch
Molniya 2-7 (1973-76A)	19 Oct 1033	63	509	39 855	

Satellite name and designation	Launch date and time GMT	Orbital inclination deg	Perigee height km	Apogee height km	Comments
Molniya 1-25 (1973-89A)	14 Nov 2038	65	454	39 197	
Molniya 1-26 (1973-97A)	30 Nov 1312	63	619	40 829	
Cosmos 614 (1973-98A)	4 Dec 1507	74	770	805	See reference [25a]; octuple launch
Cosmos 617 (1973-104A)	19 Dec 0936	74	1 336	1 436	Octuple launch
Cosmos 618 (1973-104B)	19 Dec 0936	74	1 446	1 486	Octuple launch
Cosmos 619 (1973-104C)	19 Dec 0936	74	1 423	1 493	Octuple launch
Cosmos 620 (1973-104D)	19 Dec 0936	74	1 461	1 495	Octuple launch
Cosmos 621 (1973-104E)	19 Dec 0936	74	1 410	1 485	Octuple launch
Cosmos 622 (1973-104F)	19 Dec 0936	74	1 371	1 487	Octuple launch
Cosmos 623 (1973-104G)	19 Dec 0936	74	1 389	1 487	Octuple launch
Cosmos 624 (1973-104H)	19 Dec 0936	74	1 366	1 474	Octuple launch
Molniya 2-8 (1973-106A)	25 Dec 1117	63	488	40 809	
1974					
Molniya 1-27 (1974-23A)	20 Apr 2053	63	624	40 707	
Cosmos 641 (1974-24A)	23 Apr 1410	74	1 389	1 484	Octuple launch
Cosmos 642 (1974-24B)	23 Apr 1410	74	1 321	1 483	Octuple launch
Cosmos 643 (1974-24C)	23 Apr 1410	74	1 355	1 484	Octuple launch
Cosmos 644 (1974-24D)	23 Apr 1410	74	1 336	1 484	Octuple launch
Cosmos 645 (1974-24E)	23 Apr 1410	74	1 370	1 485	Octuple launch
Cosmos 646 (1974-24F)	23 Apr 1410	74	1 405	1 487	Octuple launch
Cosmos 647 (1974-24G)	23 Apr 1410	74	1 424	1 486	Octuple launch
Cosmos 648 (1974-24H)	23 Apr 1410	74	1 440	1 490	Octuple launch
Molniya 2-9 (1974-26A)	26 Apr 1424	63	600	40 702	
Molniya 2-10 (1974-56A)	23 Jul 0126	63	604	40 726	

Satellite name and designation	Launch date and time GMT	Orbital inclination deg	Perigee height km	Apogee height km	Comments
Molniya 1-S-1 (1974-60A)	29 Jul 1200	0	35 787	35 790	First stationary Molniya satellite
Cosmos 676 (1974-71A)	11 Sep 1746	74	796	816	See reference [25a]
Cosmos 677 (1974-72A)	19 Sep 1438	74	1 399	1 469	Octuple launch
Cosmos 678 (1974-72B)	19 Sep 1438	74	1 468	1 535	Octuple launch
Cosmos 679 (1974-72C)	19 Sep 1438	74	1 468	1 513	Octuple launch
Cosmos 680 (1974-72D)	19 Sep 1438	74	1 468	1 494	Octuple launch
Cosmos 681 (1974-72E)	19 Sep 1438	74	1 468	1 474	Octuple launch
Cosmos 682 (1974-72F)	19 Sep 1438	74	1 455	1 468	Octuple launch
Cosmos 683 (1974-72G)	19 Sep 1438	74	1 436	1 469	Octuple launch
Cosmos 684 (1974-72H)	19 Sep 1438	74	1 418	1 468	Octuple launch
Molniya 1-28 (1974-81A)	24 Oct 1243	63	656	40 614	
Molniya 3-1 (1974-92A)	21 Nov 1033	63	625	40 685	
Molniya 2-11 (1974-102A)	21 Dec 0224	63	659	40 629	
1975					
Molniya 2-12 (1975-09A)	6 Feb 0448	63	634	40 660	
Cosmos 711 (1975-16A)	28 Feb 1355	74	1 462	1 496	Octuple launch
Cosmos 712 (1975-16B)	28 Feb 1355	74	1 413	1 492	Octuple launch
Cosmos 713 (1975-16C)	28 Feb 1355	74	1 398	1 490	Octuple launch
Cosmos 714 (1975-16D)	28 Feb 1355	74	1 446	1 494	Octuple launch
Cosmos 715 (1975-16E)	28 Feb 1355	74	1 470	1 508	Octuple launch
Cosmos 716 (1975-16F)	28 Feb 1355	74	1 480	1 517	Octuple launch
Cosmos 717 (1975-16G)	28 Feb 1355	74	1 481	1 538	Octuple launch
Cosmos 718 (1975-16H)	28 Feb 1355	74	1 430	1 492	Octuple launch
Molniya 3-2 (1975-29A)	14 Apr 1800	63	608	40 661	

Satellite name and designation	Launch date and time GMT	Orbital inclination deg	Perigee height km	Apogee height km	Comments
Molniya 1-29 (1975-36A)	29 Apr 1033	63	430	40 852	
Cosmos 732 (1975-45A)	28 May 0029	74	1 405	1 472	Octuple launch
Cosmos 733 (1975-45B)	28 May 0029	74	1 472	1 555	Octuple launch
Cosmos 734 (1975-45C)	28 May 0029	74	1 445	1 473	Octuple launch
Cosmos 735 (1975-45D)	28 May 0029	74	1 462	1 477	Octuple launch
Cosmos 736 (1975-45E)	28 May 0029	74	1 471	1 489	Octuple launch
Cosmos 737 (1975-45F)	28 May 0029	74	1 471	1 532	Octuple launch
Cosmos 738 (1975-45G)	28 May 0029	74	1 469	1 512	Octuple launch
Cosmos 739 (1975-45H)	28 May 0029	74	1 425	1 473	Octuple launch
Molniya 1-30 (1975-49A)	5 Jun 0141	63	435	40 857	
Molniya 2-13 (1975-63A)	8 Jul 0502	63	432	40 862	
Molniya 1-3 (1975-79A)	2 Sep 1312	63	623	40 667	
Molniya 2-14 (1975-81A)	9 Sep 0029	63	439	40 837	
Cosmos 761 (1975-86A)	17 Sep 0712	74	1 402	1 484	Octuple launch
Cosmos 762 (1975-86B)	17 Sep 0712	74	1 440	1 487	Octuple launch
Cosmos 763 (1975-86C)	17 Sep 0712	74	1 476	1 512	Octuple launch
Cosmos 764 (1975-86D)	17 Sep 0712	74	1 481	1 528	Octuple launch
Cosmos 765 (1975-86E)	17 Sep 0712	74	1 480	1 553	Octuple launch
Cosmos 766 (1975-86F)	17 Sep 0712	74	1 421	1 486	Octuple launch
Cosmos 767 (1975-86G)	17 Sep 0712	74	1 457	1 490	Octuple launch
Cosmos 768 (1975-86H)	17 Sep 0712	74	1 474	1 493	Octuple launch
Cosmos 773 (1975-94A)	30 Sep 1843	74	791	808	See reference [25a]
Molniya 3-3 (1975-105A)	14 Nov 1912	63	523	40 790	
Cosmos 783 (1975-112A)	28 Nov 0014	74	795	815	See reference [25a]

Satellite name and designation	Launch date and time GMT	Orbital inclination deg	Perigee height km	Apogee height km	Comments
Molniya 2-15 (1975-121A)	17 Dec 1117	63	431	40 821	
Statsionar 1 (1975-123A)	22 Dec 1312	0	35 800	35 800	
Molniya 3-4 (1975-125A)	27 Dec 1033	63	443	40 764	
1976					
Molniya 1-32 (1976-06A)	22 Jan 1146	63	476	39 579	
Cosmos 791 (1976-08A)	28 Jan 1033	74	1 402	1 490	Octuple launch
Cosmos 792 (1976-08B)	28 Jan 1033	74	1 436	1 494	Octuple launch
Cosmos 793 (1976-08C)	28 Jan 1033	74	1 418	1 494	Octuple launch
Cosmos 794 (1976-08D)	28 Jan 1033	74	1 452	1 497	Octuple launch
Cosmos 795 (1976-08E)	28 Jan 1033	74	1 467	1 503	Octuple launch
Cosmos 796 (1976-08F)	28 Jan 1033	74	1 474	1 518	Octuple launch
Cosmos 797 (1976-08G)	28 Jan 1033	74	1 480	1 533	Octuple launch
Cosmos 798 (1976-08H)	28 Jan 1033	74	1 481	1 557	Octuple launch
Molniya 1-33 (1976-21A)	11 Mar 1955	63	491	40 682	
Molniya 1-34 (1976-26A)	19 Mar 1938	63	416	38 882	
Molniya 3-5 (1976-41A)	12 May 1800	63	625	40 657	
Cosmos 825 (1976-54A)	15 Jun 1312	74	1 397	1 489	Octuple launch
Cosmos 826 (1976-54B)	15 Jun 1312	74	1 484	1 546	Octuple launch
Cosmos 827 (1976-54C)	15 Jun 1312	74	1 415	1 491	Octuple launch
Cosmos 828 (1976-54D)	15 Jun 1312	74	1 435	1 491	Octuple launch
Cosmos 829 (1976-54E)	15 Jun 1312	74	1 453	1 492	Octuple launch
Cosmos 830 (1976-54F)	15 Jun 1312	74	1 471	1 495	Octuple launch
Cosmos 831 (1976-54G)	15 Jun 1312	74	1 477	1 510	Octuple launch
Cosmos 832 (1976-54H)	15 Jun 1312	74	1 484	1 523	Octuple launch

Satellite name and designation	Launch date and time GMT	Orbital inclination deg	Perigee height km	Apogee height km	Comments
Cosmos 836 (1976-61A)	29 Jun 0810	74	791	818	
Cosmos 841 (1976-69A)	15 Jul 1312	74	787	808	
Molniya 1-35 (1976-74A)	23 Jul 1550	63	476	39 045	
Cosmos 853 (1976-88D)	1 Sep 0322	63	242	473	May be a failed Molniya satellite
Statsionar 1B (Raduga) (1976-92A)	11 Sep 1829	0	35 900	35 900	
Cosmos 858 (1976-98A)	29 Sep 0712	74	792	813	
Statsionar 1C (Ekran) (1976-107A)	26 Oct 1453	0	35 852	35 850	
Molniya 3-6 (1976-127A)	28 Dec 0643	63	544	39 773	

Table 4A.3. British communications satellites with possible military applications

Satellite name and designation	Launch date and time GMT	Orbital inclination deg	Perigee height km	Apogee height km	Comments
	1969				
UK/NASA Skynet 1 (1969-101A)	22 Nov 0043	2	34 702	35 838	First synchronous military defence communications satellite placed over Indian Ocean
	1970				
UK/NASA Skynet 2 (1970-62A)	19 Aug 1214	28	270	36 041	A standby satellite to Skynet 1; failed to achieve required orbit
	1974				
UK/NASA Skynet 2A (1974-02A)	17 Jan 0141	38	96	3 406	
UK/NASA Skynet 2B (1974-94A)	23 Nov 0029	2	35 784	35 794	

Table 4A.4. NATO communications satellites

Satellite name and designation	Launch date and time *GMT*	Orbital inclination *deg*	Perigee height *km*	Apogee height *km*
	1970			
US/NATO 1 (1970-21A)	20 Mar 2346	3	34 429	35 860
	1971			
US/NATO 2 (1971-09A)	3 Feb 0141	3	34 429	35 860
	1976			
US/NATO 3A (1976-35A)	22 Apr 2053	27	35 778	35 797

Table 4A.5. French communications satellites with possible military applications

Satellite name and designation	Launch date and time *GMT*	Orbital inclination *deg*	Perigee height *km*	Apogee height *km*
	1974			
Symphonie 1 (1974-101A)	19 Dec 0258	0	35 768	35 806
	1975			
Symphonie 2 (1975-77A)	27 Aug 0141	0	35 776	35 797

5. Navigation satellites

The increasing complexity of modern warfare involving a host of mobile weapon systems such as aircraft, missiles and ships, puts enormous demands on the new military navigation systems. To be able to direct operations requires exact knowledge of weapon position, velocity and direction. These requirements are being fulfilled by a variety of techniques and equipment such as Loran and Omega radio systems, navigation radar, inertial guidance and navigation satellites. Potentially, the latter technology represents a continuous worldwide navigation capability with a high degree of accuracy.

I. Some basic concepts

Basically the navigation satellite's function is to transmit, on very stable frequencies, signals that provide a constant reference frequency, a navigation message describing the satellite's position as a function of time, and timing signals. Updated navigation messages and time corrections are periodically transmitted from the ground stations to the satellite. By receiving these signals during a single pass, a navigator can calculate his position accurately.

In order to determine his position, the navigator must relate it to the known position of the satellite in orbit. In practice the position and velocity of a navigator in space are determined from simultaneous observations made by the navigator of the range and range rate (the rate at which the distance between the navigator and the satellite changes) at any given time relative to the known positions and velocities of three navigation satellites. If the centre of the Earth is the origin of the coordinate system in which the satellite positions are known, then the navigator's position is determined as the intersection of three spheres represented by the following equations [5]:

$$(x_o - x_i)^2 + (y_o - y_i)^2 + (z_o - z_i')^2 = \rho_{io}^2 \tag{5.1}$$

where
$$i = 1, 2, 3$$
x_o, y_o, z_o = Cartesian position coordinates of the observer,
x_i, y_i, z_i = known Cartesian position coordinates of the three navigation satellites,

131

and

ρ_{io} = ranges from the three navigation satellites to the navigator.
The navigator's velocity is determined from

$$\dot{\rho}_{io} = [(\bar{V}_o - \bar{V}_i)\ \bar{\rho}_{io}]/\bar{\rho}_{io} \tag{5.2}$$

where

$\bar{V}_o - \bar{V}_i = (\dot{x}_o - \dot{x}_i)\bar{i} + (\dot{y}_o - \dot{y}_i)\bar{j} + (\dot{z}_o + \dot{z}_i)\bar{k},$
$\dot{x}_o, \dot{y}_o, \dot{z}_o$ = Cartesian velocity coordinates of the observer,
$\dot{x}_i, \dot{y}_i, \dot{z}_i$ = Cartesian velocity components of the three navigation satellites,
$\dot{\rho}_{io}$ = the range rate data,

and $\bar{\rho}_{io}$ is given from Equation (5.1).

Radio signals received from a moving satellite appear higher in frequency as the satellite approaches the navigator and lower in frequency as the satellite recedes from the observer. The difference between the observed frequency and the known transmitter frequency is called the Doppler shift. The Doppler shift is a measure of relative motion of the satellite and the navigator or the relative position when the relative motions are integrated. The two unknown quantities in Equations (5.1) and (5.2) can thus be determined using this technique and the navigator's position calculated.

In the Doppler technique, the frequencies received from the satellites are compared with those generated in the navigator's equipment so that the shift in frequency due to the satellite's motion is determined. The received frequency f

Figure 5.1. The Doppler principle applied to a satellite

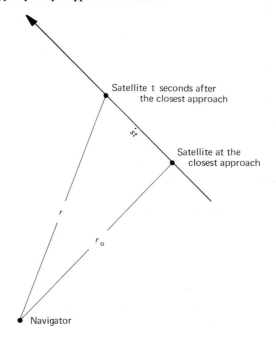

is a function of the transmitted frequency f_0, of the velocity of signal propagation and of the rate of change of the distance r between the satellite and the navigator (see Figure 5.1). These quantities are related as follows:

$$f = f_0(1 + \dot{r}/c) \tag{5.3}$$

where

$$c = \text{the velocity of light}$$

and

$$\dot{r} = \text{the radial velocity of the satellite.}$$

If the frequency f is recorded, the radial velocity may be calculated from

$$\dot{r} = c(f - f_0)/f_0. \tag{5.4}$$

When the satellite is closest to the navigator, the radial velocity is zero, since the relative velocity $\dot{s}$ is assumed to be perpendicular to the line of sight (see Figure 5.1). After a time t, the satellite will be at a distance $\dot{s}t$ from the point of closest approach and its distance from the observer is

$$r = \sqrt{(r_0^2 + \dot{s}^2 t)} \tag{5.5}$$

where
$r_0 = $ the minimum distance between the navigator and the satellite.

Squaring Equation (5.5) and differentiating it with respect to time gives the result

$$r^2 \dot{r}^2 = \dot{s}^4 t^2.$$

Substitute for r from equation (5.5) to get

$$(t/\dot{r})^2 = (t^2/\dot{s}^2) + (r_0^2/\dot{s}^4).$$

Multiply both sides of this equation by $(c/f_0)^2$ to get

$$t^2/(\dot{r}f_0/c)^2 = (c/f_0\dot{s})^2 t^2 + (cr_0/f_0\dot{s}^2)^2$$

or

$$(t/\Delta f)^2 = (c/f_0\dot{s})^2 t^2 + (cr_0/f_0\dot{s}^2)^2. \tag{5.6}$$

From Equation (5.4) it can be seen that

$$\dot{r}f_0/c = f - f_0 = \Delta f$$

where Δf is the Doppler shift which is measured. Since the instant t of the observation is also measured, the left-hand side of Equation (5.6) may be considered as the result of one observation. On the right-hand side of Equation (5.6), $(c/f_0\dot{s})^2$ of the first term and the whole of the second term may be considered constant for the observed part of the orbit which is assumed to be circular. The variable in the equation is therefore t^2.

If a graph of $(t/\Delta f)^2$ against t^2 is plotted, then the slope of the straight line obtained is given by $(c/f_o \dot{s})^2$ and the intercept by $(cr_0/f_o \dot{s}^2)^2$. From these, the minimum distance r_0 (or the range) and the relative velocity s (range rate) can be determined. Therefore, using these results and Equations (5.1) and (5.2), the navigator's position can be calculated.

The observation time is usually about 15 min and the oscillator frequency is sufficiently stable to give accurate readings. A number of other factors affect the accuracy of the system, such as: (a) the instrument and measurement noise in the form of local and satellite oscillator phase jitter, or navigator's clock error; (b) uncertainties in the geopotential model used in generating the orbit; (c) incorrectly modelled surface forces such as drag and radiation pressure acting on the satellites; (d) ephemeris rounding error (the last digit in the ephemeris is rounded); and (e) uncorrected propagation effects – ionospheric and tropospheric effects.

The root sum square (rss) of all the errors lies in the range of from 18 to 35 m. Experiments indicate that the rss range is from 27 to 37 m with a maximum error of 77 m [70].

Two other main sources of error result from refraction of the signal radiation transmitted by satellites, the greater of the two being due to the ionosphere. The wavelength of radiation passing through the ionosphere is somewhat distorted due to interaction with free electrons and ions. The amount of distortion of this wavelength is roughly inversely proportional to the square of transmitted frequency. The path length of the transmitted signal through the ionosphere changes because of the satellite's motion round the Earth. The rate of change of the path length causes an ionospheric refraction frequency shift error in the received signal. It has been shown, however, that this can be corrected if the Doppler measurements are made at two different frequencies [71]; US Transit satellites, for example, transmit coherent signals at both 150 and 400 MHz.

If only one frequency is used for transmission, for example 400 MHz, the total magnitude of errors caused by the factors enumerated above varies from very small to 200–500 m, because the density of the ionosphere varies from less dense at night to highly dense during daytime. The density also depends on sunspot activity and location with respect to the magnetic equator where the ionosphere is most dense. Using a high frequency results in a maximum error of 242 m and an rss error of 88 m [70].

The second type of error is introduced by the troposphere. As the signal radiation passes through the Earth's atmosphere, the signal wavelength is compressed because the speed of propagation is reduced. This effect is directly proportional to transmission frequency and cannot therefore be detected in the same manner as ionospheric refraction. The effect of tropospheric refraction can be reduced in two ways: in one method, the effect is compensated by knowledge of temperature, pressure and humidity; in the second method, the tropospheric effects are reduced by excluding any data obtained close to the horizon. Above 5° to 10° of elevation, the tropospheric error is many times smaller than that at the horizon.

II. The US programme

Military navigation satellites have been proposed for three altitudes: low altitude (900–2700 km), medium altitude (13 000–20 000 km) and synchronous altitude (22 000–48 000 km). At these altitudes the orbital periods of the satellites are 100–150 min, 8–12 h and 24 h, respectively. All the US navigation satellites launched up to 1974 were of the low-altitude type. The present system uses the Doppler technique to determine the navigator's position. At low altitudes the Doppler shift recorded by the navigator is more pronounced than at medium or synchronous altitudes because of the greater difference in velocity between the satellite and the navigator. In this technique the satellite transmits two types of signals. A radio wave is transmitted on two carrier frequencies – approximately 150 and 400 MHz. The satellite also transmits in code the latest information about its own orbit. The latter is recorded in the memory of the satellite's computer each time it passes over a ground station.

The US Navy was initially concerned with the development of navigation satellites under the Navy Navigation Satellite System (NNSS) programme. A

Figure 5.2. The US navigation satellite Transit IIA

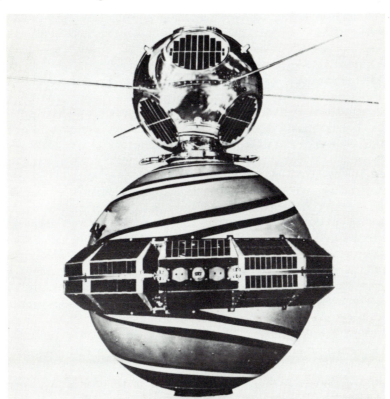

series of satellites designated Transit (Figure 5.2) were launched, beginning with Transit 1A on 17 September 1959, which was launched using a Thor rocket with an Able upper stage. This satellite was the first of its kind to use the Doppler technique for navigation. Until November 1961 Thor/Able Star rockets were used to launch Transit satellites from ETR, but at the end of 1962 Scout launchers were introduced and the satellites were launched from Vandenberg Air Force Base (WTR). These four-stage vehicles are relatively inexpensive and were used to launch small payloads on a wide variety of missions. The Transit satellite was designed basically as an aid to navigation for missile submarines, but towards the end of 1964 secrecy descended over the navigation satellite programme and the designation Transit ceased to be used.

A second set of satellites under the NNSS programme are TIMATION (Time Navigation) satellites, sometimes also referred to as Navigation Technology Satellites (NTS). Navigation using these satellites is based on measuring the time a radio signal takes to travel from the satellite to the navigator's receiving equipment. Three such experimental satellites have been launched so far, one in May 1967 (TIMATION I, 1967-53F), one in 1969 (TIMATION II, 1969-82B) and a third in 1974 (TIMATION III, 1974-54A).

A further programme called the Defense Navigation Satellite Development Program is also under consideration by the US Air Force. The proposed demonstration satellite system would consist of four satellites, the centre one at a synchronous altitude and the remaining three in inclined elliptical orbits with 24-h periods. USAF navigation technology is in some respects similar to that developed by the USN. In both systems, for example, the position of the navigator is determined by making simultaneous measurements of the distance (range) from the navigator to each of several satellites. The differences between the two systems lie in the orbital configuration and satellite altitude. The USN orbits its satellites in polar orbits at medium altitude. Only three satellites are needed to give global coverage but the navigator can only determine his position intermittently and in two dimensions. However, a total of 27 satellites would provide continuous global coverage in three dimensions, including the altitude. In contrast, the USAF plans to have a combination of satellites in geosynchronous and semisynchronous orbits. In this case a minimum of four or five satellites is needed in a regional constellation and a total of 20 satellites, five in each regional constellation, to cover the Earth. A further difference between the two systems is the fact that a USN satellite, such as TIMATION, has an accurate crystal oscillator on board so that each spacecraft operates autonomously except for periodic updating of its orbital parameters and clock from several ground stations situated in various parts of the world. In the case of the USAF programme, on the other hand, the satellites are used as intermediate transmitters of signals originating at a central ground station for each regional constellation. For determining the position of each satellite in orbit, several additional small ground terminals would be required for each regional constellation.

Although the desirability of a joint services programme for the development of a satellite navigation system, which would provide accurate navigation capability to ground-based, airborne or shipborne weapon systems, had been recognized for some time, it was not until mid-1974 that such a programme was created. A new system called the Global Positioning System or the Navstar was planned.

The Navstar system will consist of 24 satellites grouped equally into three rings situated in circular orbits, at altitudes of about 20 000 km with 12-h periods, and at orbital inclinations of 63°. It is expected that with such a system, a navigator will be able to obtain continuous position fixes in three dimensions to within about 10 m and will be able to determine his velocity to within about 6 cm/s [72]. The system is designed particularly with weapon-delivery systems in mind, and will, for example, be able to navigate ICBMs to their targets very accurately. It is envisaged for use with nuclear weapons and for synchronizing the automated battlefield [73].

The Navstar system is expected to become operational in 1984. A Navigation Technology Satellite, NTS 1, was launched on 14 July 1974 from ETR to test techniques under consideration for use in the Navstar system. The satellite was equipped to transmit on two frequencies, 335 and 1580 MHz, so that an evaluation of dual-frequency operation could be made, particularly from the point of view of improving navigational accuracy. The satellite also carried a quartz crystal oscillator and a rubidium atomic clock [74]. However, tests of these clocks and other experiments have had limited success because of satellite stabilization problems [75]. In early 1977, NTS 2 was to have been launched to test advanced atomic and crystal frequency standards for use in the Navstar system [76] but the programme was delayed [77]. However, NTS-2 was launched on 23 June 1977 and some of the basic navigation tests were successfully carried out. If the new plans are fulfilled, then the evaluation will begin in early 1978 with at least four satellites in orbit [77]. Future developments include caesium atomic clocks and hardened satellites to minimize vulnerability to both nuclear and non-nuclear attacks. The possibility of using Navstar for navigating other spacecraft in geosynchronous or elliptical orbits is also being considered [78]. The orbital characteristics of these satellites are shown in Table 5A.1.

III. The Soviet programme

Although the Soviet Union has publicized the operation of a navigation satellite system, no details about the satellites have been given. The navigation satellites fall within the Cosmos series which contains virtually all the numerous military satellites as well as a number of experimental and scientific satellites. However, analyses of satellite orbital data and monitoring of their telemetry signals have identified satellites which are probably used for navigation purposes. Moreover, identification of such satellites is particularly facilitated if a group of satellites with similar orbital parameters also have a geometrical relationship which allows complete global coverage.

The Kettering Group calculated the values of the right ascension of the ascending node of satellites with orbital inclinations of about 74°. These satellites were chosen because they had orbital characteristics similar to those of the US Transit navigation satellites. The Group showed that the satellites (with 74° inclinations and 105-min periods) launched during 1970–72 had orbital planes spaced at 120° intervals [7].

On 16 August 1972, Cosmos 514 was launched into an orbit with very similar characteristics to those described above, except that its orbital inclination was 82.95°. This new set of satellites provides the same kind of global coverage as those at 74° inclinations but their orbital planes are now spaced at 60° rather than 120° intervals [79].

Analyses of the right ascension of ascending nodes show that Cosmos 475 and 489 replaced Cosmos 385 and 422 at intervals of about one year, indicating that this is the useful lifetime of the payload. Similarly, Cosmos 627 and 689 replaced Cosmos 514 and 574, respectively, and Cosmos 586 was replaced by Cosmos 628 and 663. These latter satellites formed a three-satellite navigation system. In attempting to identify the radio transmissions from these satellites, it was found that the satellites transmit at 150 and 400 MHz, frequencies also used by US navigation satellites [79].

Cosmos 778 was launched on 4 November 1975. The orbital plane of this satellite and that of Cosmos 726 were 30° apart. This was the beginning of the new set in which the orbital spacing of the navigation satellites is 30°. Cosmos 789, launched on 20 January 1976, also belonged to this new system of satellites [25b].

The orbital characteristics of all the navigation satellites are given in Table 5A.2.

Appendix 5A

Tables of navigation satellites

For abbreviations, acronyms and conventions, see page xv.

The designation of each satellite is recognized internationally and is given by the World Warning Agency on behalf of the Committee on Space Research.

More detailed tables of US and Soviet navigation satellites are to be found in *World Armaments and Disarmament, SIPRI Yearbook 1977* (pp. 139–41, 158–59).

Table 5A.1. US navigation satellites

Satellite name and designation	Launch date and time *GMT*	Orbital inclination *deg*	Perigee height *km*	Apogee height *km*	Comments
	1959				
ARPA Transit 1A —	17 Sep ..	—	—	—	Failed to orbit
	1960				
ARPA Transit 1B (1960-γ2)	13 Apr 1200	51	373	748	
USN Transit 2A (1960-η1)	22 Jun 0600	67	628	1 047	
USN Transit 3A —	30 Nov ..	—	—	—	Failed to orbit
	1961				
USN Transit 3B (1961-η1)	22 Feb 0350	28	167	1 002	
USN Transit 4A (1961-σ1)	29 Jun 0419	67	881	998	
USN Transit 4B (1961-αη1)	15 Nov 2219	32	956	1 104	

Satellite name and designation	Launch date and time GMT	Orbital inclination deg	Perigee height km	Apogee height km	Comments
	1962				
USN Transit 5A (1962-βψ1)	19 Dec 0126	91	698	723	
	1963				
USAF —	.. ..	—	—	—	Failed to orbit
USN Transit (1963-22A)	16 Jun 0155	90	724	757	
USAF/USN Transit 5B? (1963-38B)	28 Sep 2010	90	1 075	1 127	
USAF/USN Transit (1963-49B)	5 Dec 2150	90	1 067	1 112	
	1964				
USN —	21 Apr ..	—	—	—	Failed to orbit
USN Transit (1964-26A)	4 Jun 0350	90	854	956	
USAF/USN Transit ? (1964-63B)	6 Oct 1702	90	1 055	1 085	
USAF/USN Transit (1964-83D)	13 Dec 0014	90	1 025	1 084	
	1965				
USAF/USN Transit ? (1965-17A)	11 Mar 1341	90	211	890	
USN Transit ? (1965-48A)	24 Jun 2234	90	1 024	1 144	
USN Transit (1965-65F)	13 Aug 2248	90	1 089	1 194	
USN Transit ? (1965-109A)	22 Dec 0434	89	909	1 080	
	1966				
USN Transit ? (1966-05A)	28 Jan 1702	90	861	1 217	

140

Satellite name and designation	Launch date and time GMT	Orbital inclination deg	Perigee height km	Apogee height km	Comments
USN Transit ? (1966-24A)	26 Mar 0336	90	891	1 128	
USN Transit ? (1966-41A)	19 May 0224	90	863	980	
USN Transit ? (1966-76A)	18 Aug 0224	89	1 056	1 101	
	1967				
USN — (1967-34A)	14 Apr 0322	90	1 053	1 083	
USN — (1967-48A)	18 May 0907	90	1 074	1 105	
USN — (1967-92A)	25 Sep 0824	90	1 041	1 116	
	1968				
USAF — (1968-12A)	2 Mar 0350	90	1 035	1 139	
	1970				
USN Navy Navigation Satellite 19 (1970-67A)	27 Aug 1326	90	955	1 221	
	1972				
USAF TRIAD 1 (1972-69A)	2 Sep 1746	90	716	863	
	1973				
USN Navy Navigation Satellite 20 (1973-81A)	30 Oct 0043	90	895	1 149	
	1974				
USAF NTS 1 (TIMATION 3) (1974-54A)	14 Jul 0517	125	13 445	13 767	

Satellite name and designation	Launch date and time GMT	Orbital inclination deg	Perigee height km	Apogee height km	Comments
1975					
USAF TRIAD 2 (TIP 2) (1975-99A)	12 Oct 0643	91	362	705	
1976					
USAF TIP 3 (1976-89A)	1 Sep 2107	90	348	789	

5A.2. Soviet navigation satellites

Satellite name and designation	Launch date and time GMT	Orbital inclination deg	Perigee height km	Apogee height km	Comments
1970					
Cosmos 385 (1970-108A)	12 Dec 1258	74	978	986	
1971					
Cosmos 422 (1971-46A)	22 May 0043	74	988	1 010	
Cosmos 465 (1971-111A)	15 Dec 0434	74	970	1 012	
1972					
Cosmos 475 (1972-09A)	25 Feb 0936	74	970	1 000	
Cosmos 489 (1972-35A)	6 May 1117	74	969	1 002	
Cosmos 514 (1972-62A)	16 Aug 1522	83	958	975	
1973					
Cosmos 574 (1973-42A)	20 Jun 0614	83	985	1 014	
Cosmos 586 (1973-65A)	14 Sep 0029	83	971	1 009	
Cosmos 627 (1973-109A)	29 Dec 0405	83	974	1 019	
1974					
Cosmos 628 (1974-01A)	17 Jan 1005	83	958	1 016	
Cosmos 663 (1974-48A)	27 Jun 1536	83	972	1 007	

Satellite name and designation	Launch date and time GMT	Orbital inclination deg	Perigee height km	Apogee height km	Comments
Cosmos 689 (1974-791A)	18 Oct 2234	83	981	1 017	
Cosmos 700 (1974-105A)	26 Dec 1200	83	966	999	
1975					
Cosmos 726 (1975-28A)	11 Apr 0755	83	956	996	
Cosmos 729 (1975-34A)	22 Apr 2107	83	980	1 011	
Cosmos 755 (1975-74A)	14 Aug 1326	83	974	1 013	
Cosmos 778 (1975-103A)	4 Nov 1005	83	978	1 004	
1976					
Cosmos 789 (1976-05A)	20 Jan 1702	83	975	1 016	
Cosmos 800 (1976-11A)	3 Feb 0810	83	984	1 015	
Cosmos 823 (1976-51A)	2 Jun 2234	83	980	1 011	
Cosmos 842 (1976-70A)	21 Jul 1019	83	972	1 011	
Cosmos 846 (1976-78A)	29 Jul 1955	83	954	1 015	
Cosmos 864 (1976-108A)	29 Oct 1243	83	966	1 011	
Cosmos 883 (1976-122A)	15 Dec 1355	83	961	1 012	
Cosmos 887 (1976-128A)	28 Dec 0735	83	954	1 018	

6. Meteorological satellites

The traditional method of weather prediction relies on meteorological data collected by a worldwide network of observers. While an enormous quantity of information can be obtained in this way, it is by no means comprehensive, since little day-to-day meteorological information is obtained from large areas of the Earth such as the oceans and the polar regions. Meteorological satellites now fill this gap and give meteorologists a view of weather patterns over the entire Earth. In addition to ascertaining cloud coverage, the infra-red sensors in the satellites can estimate temperature patterns at various altitudes within the Earth's atmosphere. Wind direction can also be determined from such satellites.

These meteorological data are, of course, not only useful in the civilian field; they have widespread military applications as well. Knowledge of cloud formation and movements, for example, is of utmost importance in the photography by reconnaissance satellites of targets of military interest, and in bombing missions. Moreover, a more sinister aspect of the development of meteorological satellites lies in the fact that once man eventually understands the mechanics of weather formation, his military mind may acquire the ability to control weather and use it for hostile purposes. The part that space technology plays at present is that of providing a better understanding of the Earth's atmosphere.

I. Satellite orbits

The US and Soviet weather, or meteorological, satellites are launched in quite different orbits from those considered above. A circular orbit is usually used with satellite altitudes of between 500 and 1500 km – altitudes low enough for cloud details to be seen and high enough for wide views which overlap on successive revolutions to be photographed. US satellites are orbited in near-polar orbits so that each satellite can obtain a complete picture of the globe every 24 h.

Special orbital inclinations are often chosen, for example for the US Nimbus satellite, to ensure that the satellites pass over the same area at the same time of day throughout their orbital lifetime so that the photographs taken of weather conditions always refer to the same local time. Such an orbit is achieved if the precession turns the orbital plane anticlockwise as seen from the north at the same rate at which the Earth rotates around the Sun. This means that the orbital plane remains stationary with respect to the Sun. An orbital inclination of about 100° produces the required precession of 0.98°/day, an orbital inclination not very far from a polar orbit.

II. The US programme

The US Army, Navy and industry began to study weather satellite technology in the early 1950s – in particular, the Radio Corporation of America (RCA) applied to meteorology the experience it had gained studying television-equipped satellites for the Air Force. But it was not until early 1973 that the

Figure 6.1. A US defence meteorological satellite for use by the USAF

USAF first acknowledged that it had been operating its own weather satellites (Figure 6.1) [80]. In the early 1960s, the USAF relied upon cloud-cover photographs obtained from the civilian RCA-built Tiros satellite. However, civilian and military meteorological requirements are very different; the military require high-resolution cloud-cover photographs over specific geographical regions, whereas the civilian requirement is for low resolution but for wider area coverage. The military therefore embarked on a separate weather satellite programme. Very little is known about this programme, designated Program 417 [81], but its requirement was to develop a weather reconnaissance satellite which would determine when a particular region on Earth is free of cloud cover so that the surveillance satellite could photograph the region. It has been reported that the first experimental military weather satellite built by RCA was launched into polar orbit from WTR on 18 January 1965 using the Thor/Altair rocket [80]. The satellite weighed 73 kg and carried a payload of vidicon cameras. More advanced weather satellites have subsequently been launched from WTR. On average, two or three satellites have been launched each year to secure global coverage for military forces around the world.

Figure 6.2. Photograph of the weather pattern on the Earth's disc, taken from the first US synchronous meteorological satellite SMS-1 on 28 May 1974, with a resolution of 0.9 km

The satellites, communications links and terminal equipment have been considerably improved in recent years. USAF weather satellites are now able to provide very high-resolution, visible and infra-red photographs of cloud cover of all parts of the world twice daily. The resolution of the satellite sensors is between 0.6 and about 4 km (Figure 6.2) [80]. Measurements of vertical profiles of atmospheric temperature are continuously being made so that adequate global distribution of upper-air temperature data is becoming available. This will potentially improve and expand the capabilities of military weather services.

There are two large, fixed, weather satellite-data receiving terminals in the United States: one at Fairchild Air Force Base (Washington) and the other at Loring Air Force Base (Maine). Weather data are first stored on tape recorders aboard the satellite and then transmitted to remote overseas ground stations. The data are then relayed to the two ground stations via relay satellites. From these stations the data are immediately transmitted to the USAF Global Weather Center at Offutt Air Force Base (Nebraska). The photographic images can be converted to digital data and processed by computers, so that observations, analyses and forecasts can be generated automatically. Final output is distributed through the USAF Automated Weather Network directly to military installations around the world. The Air Force also has tactical air transportable data-acquisition terminals in vans which can be flown to any part of the world and set up within hours to receive data from the satellites. The USAF has also developed a second-generation satellite – the Block 5D Integrated Spacecraft – to provide increased meteorological capability. This new satellite is being designed and built by RCA.

Since 1973, the USN has also participated in the military weather-satellite programme. Under this USAF/USN programme, tests – consisting of installing two data-receiving and -processing terminals on two ships – were conducted during 1972 and 1973 to determine the ability of ships at sea to receive real-time weather data from satellites in polar and Sun-synchronous orbits [82]. The USN is planning to install a complex data-processing centre at its Fleet Numerical Weather Center (Monterey, California) where all the weather data and photographs will be processed. At present the Navy can receive, at Monterey, data directly from the USAF satellites or indirectly from the USAF's Global Weather Center [83].

It is possible that the military may also use the two Synchronous Meteorological Satellites (SMS 1 and SMS 2) in geostationary orbits over the equator; they provide cloud-cover photographs at 30-min intervals on a continuous basis. The planned operational lifetime of the satellites is five years.

The orbital characteristics of these satellites are given in Table 6A.1.

III. The Soviet programme

Although two or three meteorological satellites are launched per year by the United States, many more Soviet meteorological satellites are launched because of their relatively short active lifetimes of about six months. The Soviet meteorological satellite programme began in 1963 with component testing in some of the satellites in the Cosmos series. For example, Cosmos 45 (launched in 1964) and Cosmos 65 and 92 (launched in 1965) were recoverable satellites and had meteorological missions. Initially experimental meteorological satellites, Cosmos 14 and 23, were launched from Kapustin Yar using the Sandal intermediate-range ballistic missile (IRBM) with an upper stage (B-1). The first known meteorological satellite, Cosmos 122, was launched from Tyuratam on 25 June 1966, witnessed by President de Gaulle. Cosmos 122 could observe the dark side of the Earth and photograph it using infra-red sensors [84]. This was the last meteorological satellite to be launched into a 65° orbit from Tyuratam. Cosmos 144 (Figure 6.3) was the first meteorological satellite to be launched from Plesetsk at an orbital inclination of 81°.

Figure 6.3. The Soviet meteorological satellite Cosmos 144

The third Molniya 1 communications satellite also transmitted photographs of cloud cover over the Earth. The routine use of satellites for monitoring weather began in 1969. The new series of satellites, called Meteors, carried equipment providing photographs with higher resolution than those obtained from the US Tiros satellites. However, the coverage was not as complete as that obtained by the US satellite [85]. A series of new improved Meteor satellites, Meteor 2, will replace the Meteor 1s. The first of these, a test Meteor 2-1, was launched on 11 July 1975. Meteor 2 satellites will carry improved visible and infra-red scanning radiometers for imagery and temperature measurements. It has been reported that the image resolution from these satellites will be com-

148

parable to that obtained by US weather satellites. The new system will also be able to transmit automatically photographs of weather conditions on Earth so that weather imagery will be received by Soviet ground- or ship-based stations around the world. In the case of the Meteor 1 satellites, the weather imagery has been transmitted to only three ground stations in the Soviet Union [86].

At present there are three Meteor ground receiving stations in the Soviet Union: one, located at Obninsk, southwest of Moscow, is only a data relay station. From this station, the satellite imagery is transmitted to Moscow using a microwave system. The data are then analysed in Moscow. The two other stations are located at Novosibirsk in central Siberia and at Khavarovsk in eastern Siberia, but they have no microwave relay system. Meteor satellite data are therefore analysed locally and then transmitted to Moscow by narrow-band communications.

Although the image quality of the photographs of weather conditions and the transmission of data have improved in the new Meteor 2 satellite, the problem of Earth coverage has still not been overcome. The Soviet Union must have at least two or three satellites in orbit at a time to obtain full daily coverage of the Earth. At present, two weather satellites are maintained in orbit at a time.

In addition to the Meteor 2 satellites, the Soviet Union has developed a synchronous weather satellite, the first of which was launched on 29 June 1977 [85].

The early Cosmos weather satellites and the Meteor satellites have been launched using the Vostok type of standard launch vehicle (A-1). Meteor satellites have been launched from Plesetsk at an orbital inclination of 81°. The orbital characteristics of these satellites are given in Table 6A.2.

IV. The British and French programmes

In addition to the interest shown by the UK and France in reconnaissance and communications satellites, they have each launched a meteorological satellite (see Tables 6A.3 and 6A.4). The first British meteorological satellite, Prospero, was launched in 1971, using a Black Arrow launch vehicle. France's first meteorological satellite, Eole, was also launched in 1971 but by the United States, using the Scout rocket.

Appendix 6A

Tables of meteorological satellites

For abbreviations, acronyms and conventions, see page xv.

The designation of each satellite is recognized internationally and is given by the World Warning Agency on behalf of the Committee on Space Research.

More detailed tables of meteorological satellites are to be found in *World Armaments and Disarmament, SIPRI Yearbook 1977* (pp. 148–52, 168–71, 175–76).

Table 6A.1. US meteorological satellites

Satellite name and designation	Launch date and time GMT	Orbital inclination deg	Perigee height km	Apogee height km	Comments
1960					
NASA Tiros 1 (1960-β2)	1 Apr 1146	48	693	750	
NASA Tiros 2 (1960-π1)	23 Nov 1117	49	619	732	
1961					
NASA Tiros 3 (1961-ρ1)	12 Jul 1019	48	735	820	
1962					
NASA Tiros 4 (1962-β1)	8 Feb 1229	48	712	840	
USAF —	23 May ..	—	—	—	Failed to orbit
NASA Tiros 5 (1962-$\alpha\alpha$1)	19 Jun 1214	58	588	974	
USAF (1962-α01)	23 Aug 1146	99	620	858	
NASA Tiros 6 (1962-$\alpha\psi$1)	18 Sep 0853	58	686	713	
1963					
USAF (1963-5A)	19 Feb 0434	101	505	791	
NASA Tiros 7 (1963-24A)	19 Jun 0950	58	621	649	
USAF —	27 Sep ..	—	—	—	Failed to orbit

Satellite name and designation	Launch date and time GMT	Orbital inclination deg	Perigee height km	Apogee height km	Comments
NASA Tiros 8 (1963-54A)	21 Dec 0922	59	691	765	
1964					
USAF (1964-2B)	19 Jan 1048	99	801	830	
USAF (1964-2C)	19 Jan 1048	99	811	825	
USAF (1964-31A)	18 Jun 0448	100	828	842	
USAF (1964-31B)	18 Jun 0448	100	828	842	
NASA Nimbus 1 (1964-52A)	28 Aug 0755	99	429	937	
1965					
USAF (1965-3A)	19 Jan 0502	99	471	822	
NASA Tiros 9 (1965-04A)	22 Jan 0755	96	705	2 582	
USAF (1965-21A)	18 Mar 0448	99	525	764	
USAF (1965-38A)	20 May 1634	99	567	953	
NASA Tiros 10 (1965-51A)	2 Jul 0405	99	751	837	
USAF (1965-72A)	10 Sep 0448	99	649	1 054	
1966					
USAF —	6 Jan ..	—	—	—	Failed to orbit
ESSA 1 (Tiros 11) (1966-08A)	3 Feb 0735	98	702	845	
ESSA 2 (1966-16A)	28 Feb 1355	101	1 356	1 418	
USAF (1966-26A)	31 Mar 0434	99	634	933	
NASA Nimbus 2 (1966-40A)	15 May 0755	100	1 103	1 179	
USAF (1966-82A)	16 Sep 0434	99	705	891	
ESSA 3 (1966-87A)	2 Oct 1033	101	1 383	1 493	
1967					
ESSA 4 (1967-06A)	26 Jan 1731	102	1 328	1 443	

Satellite name and designation	Launch date and time GMT	Orbital inclination deg	Perigee height km	Apogee height km	Comments
USAF (1967-10A)	8 Feb 0755	99	796	868	
ESSA 5 (1967-36A)	20 Apr 1117	102	1 361	1 423	
USAF (1967-80A)	23 Aug 0448	99	834	892	
USAF (1967-96A)	11 Oct 0755	99	667	866	
ESSA 6 (1967-114A)	10 Nov 1800	102	1 410	1 488	
1968					
NASA/USA Nimbus B/Secor 10	18 May ..	—	—	—	Failed to orbit
USAF (1968-42A)	23 May 0434	99	817	904	
ESSA 7 (Tiros 17) (1968-69A)	16 Aug 1131	102	1 432	1 476	
USAF (1968-92A)	23 Oct 0434	99	797	855	
ESSA 8 (1968-114A)	15 Dec 1717	102	1 410	1 473	
1969					
ESSA 9 (1969-16A)	26 Feb 0735	102	1 427	1 508	
NASA Nimbus 3 (1969-37A)	14 Apr 0755	100	1 075	1 135	
USAF (1969-62A)	23 Jul 0434	99	788	856	
1970					
NASA ITOS 1 (1970-08A)	23 Jan 1131	102	1 436	1 482	
USAF (1970-12A)	11 Feb 0838	99	773	874	
NASA Nimbus 4 (1970-25A)	8 Apr 0824	100	1 095	1 100	
USAF (1970-70A)	3 Sep 0838	99	764	874	
NASA NOAA 1 (ITOS) (1970-106A)	11 Dec 1131	102	1 429	1 473	
1971					
USAF (1971-12A)	17 Feb 0350	99	763	833	
USAF (1971-87A)	14 Oct 0936	99	796	877	

152

Satellite name and designation	Launch date and time GMT	Orbital inclination deg	Perigee height km	Apogee height km	Comments
ITOS B (1971-91A)	21 Oct 1507	—	—	—	Failed to orbit
1972					
USAF (1972-18A)	24 Mar 0853	99	803	885	
NASA NOAA 2 (1972-82A)	15 Oct 1717	102	1 451	1 458	
USAF (1972-89A)	9 Nov 0502	99	813	872	
NASA Nimbus 5 (1972-97A)	11 Dec 0755	100	1 089	1 102	
1973					
NASA ITOS-E	16 Jul ..	—	—	—	Failed to orbit
USAF (1973-54A)	17 Aug 0448	99	811	852	
NASA NOAA 3 (1973-86A)	6 Nov 1702	102	1 500	1 509	
1974					
USAF (1974-15A)	16 Mar 0810	99	782	877	
NASA SMS 1 (1974-33A)	17 May 0936	2	35 741	35 830	
USAF (1974-63A)	9 Aug 0322	99	806	875	
NASA NOAA 4 (1974-89A)	15 Nov 1717	102	1 447	1 462	
1975					
NASA SMS 2 (1975-11A)	6 Feb 2248	1	35 680	36 685	
USAF DMSP (1975-43A)	24 May 0322	99	813	892	
NASA Nimbus 6 (1975-52A)	12 Jun 0810	100	1 092	1 104	
NASA GOES 1 (SMS-3) —	16 Oct 2234	1	35 770	35 770	
1976					
USAF (1976-16A)	19 Feb 0755	99	90	355	
NASA NOAA 5 (1976-77A)	29 Jul 1702	102	1 509	1 522	
USAF AMS 1 (1976-91A)	11 Sep 0810	99	818	848	

Table 6A.2. Soviet meteorological satellites

Satellite name and designation	Launch date and time *GMT*	Orbital inclination *deg*	Perigee height *km*	Apogee height *km*	Comments
1963					
Cosmos 14 (1963-10A)	13 Apr 1102	49	252	499	Test of meteorological instruments
Cosmos 23 (1963-50A)	13 Dec 1355	49	240	613	Test of meteorological instruments
1964					
Cosmos 44 (1964-53A)	28 Aug 1619	65	615	857	Precursor to meteorological satellite
Cosmos 45 (1964-55A)	13 Sep 0950	65	207	313	Observation and meteorological test; satellite recovered
1965					
Cosmos 58 (1965-14A)	26 Feb 0502	65	563	647	Precursor to meteorological satellite
Cosmos 65 (1965-29A)	17 Apr 0950	65	207	319	Observation and meteorological test; satellite recovered
Cosmos 92 (1965-83A)	16 Oct 0810	65	201	334	Observation and meteorological test; satellite recovered
Cosmos 100 (1965-106A)	17 Dec 0224	65	630	658	Precursor to meteorological satellite
1966					
Cosmos 118 (1966-38A)	11 May 1410	65	587	657	Precursor to meteorological satellite
Cosmos 122 (1966-57A)	25 Jun 1019	65	550	690	
1967					
Cosmos 144 (1967-18A)	28 Feb 1438	81	574	644	
Cosmos 149 (1967-24A)	21 Mar 1005	49	245	285	Experimental meteorological satellite
Cosmos 156 (1967-39A)	27 Apr 1243	81	593	635	
Cosmos 184 (1967-102A)	24 Oct 2324	81	600	638	
1968					
Cosmos 206 (1968-19A)	14 Mar 0936	81	598	640	
Cosmos 226 (1968-49A)	12 Jun 1312	81	579	639	

Satellite name and designation	Launch date and time *GMT*	Orbital inclination *deg*	Perigee height *km*	Apogee height *km*	Comments
1969					
Meteor 1 (1969-29A)	26 Mar 1229	81	633	687	
Meteor 2 (1969-84A)	6 Oct 0141	81	613	681	
1970					
Cosmos 320 (1970-05A)	16 Jan 1102	49	247	326	Experimental meteorological satellite
Meteor 3 (1970-19A)	17 Mar 1117	81	537	635	
Meteor 4 (1970-37A)	28 Apr 1048	81	625	710	
Meteor 5 (1970-47A)	23 Jun 1424	81	831	888	
Meteor 6 (1970-85A)	15 Oct 1131	81	626	648	
Cosmos 389 (1970-113A)	18 Dec 1619	81	642	687	
1971					
Meteor 7 (1971-03A)	20 Jan 1133	81	629	656	
Meteor 8 (1971-31A)	17 Apr 1146	81	610	633	
Meteor 9 (1971-59A)	16 Jul 0141	81	614	642	
Meteor 10 (1971-120A)	29 Dec 1048	81	878	889	
1972					
Cosmos 476 (1972-11A)	1 Mar 1117	81	617	633	
Meteor 11 (1972-22A)	30 Mar 1410	81	868	891	
Meteor 12 (1972-49A)	30 Jun 1858	81	889	905	
Meteor 13 (1972-85A)	26 Oct 2248	81	867	891	
Cosmos 542 (1972-106A)	28 Dec 1102	81	527	641	
1973					
Meteor 14 (1973-15A)	20 Mar 1117	81	873	892	
Meteor 15 (1973-34A)	29 May 1039	81	853	896	

Satellite name and designation	Launch date and time *GMT*	Orbital inclination *deg*	Perigee height *km*	Apogee height *km*	Comments
Cosmos 604 (1973-80A)	29 Oct 1410	81	615	631	
1974					
Meteor 16 (1974-11A)	5 Mar 1146	81	832	894	
Meteor 17 (1974-25A)	24 Apr 1200	81	865	894	
Meteor 18 (1974-52A)	9 Jul 1438	81	865	893	
Cosmos 673 (1974-66A)	16 Aug 0350	81	607	637	
Meteor 19 (1974-83A)	28 Oct 1019	81	843	907	
Meteor 20 (1974-99A)	17 Dec 1146	81	842	897	
1975					
Meteor 21 (1975-23A)	1 Apr 1229	81	867	893	
Cosmos 744 (1975-56A)	20 Jun 0658	81	602	635	
Meteor 2-01 (1975-64A)	11 Jul 0419	81	858	891	Experimental meteorological satellite
Cosmos 756 (1975-76A)	22 Aug 0210	81	622	634	
Meteor 22 (1975-87A)	18 Sep 0029	81	838	901	
Meteor 23 (1975-124A)	25 Dec 1858	81	842	902	
1976					
Cosmos 808 (1976-24A)	16 Mar 1731	81	602	634	
Meteor 24 (1976-32A)	7 Apr 1312	81	843	893	
Meteor 25 (1976-43A)	15 May 1341	81	846	895	
Cosmos 851 (1976-85A)	27 Aug 1438	81	568	637	
Meteor 26 (1976-102A)	15 Oct 2324	81	857	892	

156

Table 6A.3. British meteorological satellites with possible military applications

Satellite name and designation	Launch date and time GMT	Orbital inclination deg	Perigee height km	Apogee height km
Prospero (1971-93A)	**1971** 28 Oct 0141	82	547	1 582

Table 6A.4. French meteorological satellites with possible military applications

Satellite name and designation	Launch date and time GMT	Orbital inclination deg	Perigee height km	Apogee height km
Peole 1 (1970-109A)	**1970** 12 Dec 1258	15	517	747
Eole 1 (1971-71A)	**1971** 16 Aug 1843	50	677	904

7. Geodetic satellites

Geodesy is the branch of applied mathematics that deals with the shape of the Earth, its gravitational field and the exact positions of various points on the Earth's surface. An accurate knowledge of the shape of the Earth and of the precise whereabouts of points on the Earth is obviously essential for mapping purposes. The Earth's gravitational field is far from uniform since large sections of the Earth's crust have different densities. If the effects of the Earth's shape and its non-uniform gravitational field are neglected, then considerable errors may be introduced in the computations of trajectories and in the inertial guidance systems of missiles and aircraft.

Although the astrogeodetic method can connect all points of a land mass on to a consistent geodetic system of points, it cannot span the oceans unless the land masses are close together. Geodetic satellites, on the other hand, are able to provide greater accuracy over a much wider area. By using maps with grids accurately to locate specific places, and by obtaining knowledge of the Earth's gravitational field through satellites, the military are able to gain a more accurate cartographical picture, an essential requirement, for example, in the development of long-range ballistic missiles.

The orbital path of a satellite is an imperfect geometrical shape. The satellite weaves sideways and up and down as it moves in its orbit. These small orbital perturbations, ranging from a few centimetres to several metres can be detected by Earth-based tracking stations and are measured to determine the shape and distribution of the Earth's gravitational field.

The positions of places located at great distances from each other on the surface of the Earth can be determined by satellites in one of the following three ways. In one method, known as the geometric-optical satellite system, the moving satellite is photographed against a star background simultaneously from two ground stations A and B (Figure 7.1(a)). This fixes the pair of directions AS and BS and a plane which contains these directions and an as yet unknown straight line AB. Similar planes containing the line AB will be defined for many satellite positions so that the direction from a known station A to an unknown station B can be computed. By repeating this determination of directions from known to unknown stations, a worldwide network of stations can be formed, comparable to a huge triangulation net.

In a second method, the position of a satellite is determined by measuring three simultaneous distances from three known ground stations, A_1, A_2 and A_3 (see Figure 7.1(b)). After three satellite positions S_1, S_2 and S_3 have been determined, an unknown ground station B is fixed in relation to them and its

Figure 7.1. Methods of connecting points on land masses using satellites

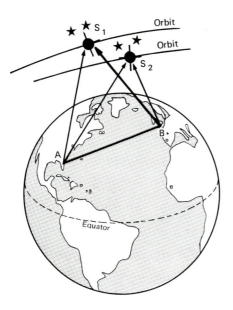

(a)   Determination of direction using satellites

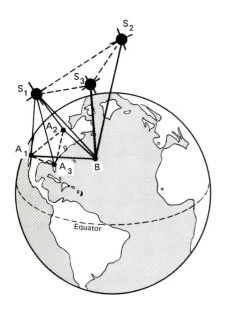

(b)   Determination of distance using satellites

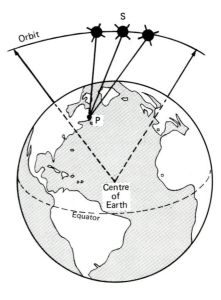

(c)   Determination of geocentric positions using satellites (Doppler method)

159

position in relation to the known ground stations can be computed. This is comparable to triangulation in three dimensions.

In the third method – a dynamic satellite system – the orbit of the satellite is determined first from equations of motion which relates the satellite's position to the centre of the Earth and include the orbit's perturbations due to an estimate of the Earth's irregular gravitational field. One type of dynamic satellite system is called the Doppler system. The Doppler satellite is tracked from a ground station P (see Figure 7.1(c)), and its nearest distance is deduced from the Doppler effect of its approach and departure. By tracking several satellites, the position of the ground station P is linked to their orbits and in turn to the centre of the Earth, yielding geocentric coordinates. If two ground points are thus linked to the same geocentric coordinate system, their positions relative to each other can be computed.

I. Satellite orbits

Basically there are two types of geodetic satellite systems – geometric and dynamic – which measure the Earth's shape, its gravitational field and positions of points on the Earth relative to each other.

A satellite used for these purposes should have a minimum perigee height of 700–1000 km so that the effects of atmospheric drag are minimized. The orbital inclination should be such that observations at higher latitudes are also possible. An inclination of 55°–70° would make the satellite accessible to most areas of interest. If, however, a satellite is used to measure the Earth's gravitational field, then various orbital inclinations are necessary.

The eccentricity of the satellite's orbit should be 0.05 or less so that geometrical problems become simpler. Such an eccentricity would ensure an almost circular orbit which would keep the satellite within an accurate observing range. For dynamical applications, larger eccentricity, of the order of 0.2, is desirable to produce a measurable motion of the perigee, but not too large to give appreciable perturbations to the major semi-axis of the orbit due to the effects of higher harmonics [87].

II. The US programme

It should be pointed out that it is possible to use almost any satellite for unsophisticated geodetic measurements. For example, the most useful of the US non-geodetic satellites were the Echo communications satellites which were easy to see with optical instruments.

160

Figure 7.2. Photograph of the first successful operational test of the US geodetic satellite ANNA's flashing light (shown within the five circles)

The early attempts in the United States in 1958 and 1959 to launch specific geodetic satellites were unsuccessful (see Table 7A.1), but on 31 October 1962, ANNA 1B was successfully orbited after ANNA 1A had failed. Pageos was another balloon satellite orbited by NASA in 1966 specifically for geodetic work.

To help make truly simultaneous observations, flashing lights were installed on several geodetic satellites. The lights flash in coded sequences so that widely separated stations can compare time-exposure photographs taken against a background of fixed stars. The DoD satellite ANNA 1B carried the first optical beacon into orbit in 1962 (see Figure 7.2). Geos A (Explorer 29), launched on 6 November by NASA, also carried a flashing light.

161

Radio beacons of various types are carried by many satellites to aid in tracking them. NASA's Explorer 22 and 27 satellites carried both laser reflectors for tracking as well as radio beacons which combined tracking functions with signals for ionospheric research. Geos B (Explorer 36), launched on 11 January 1968, carried, in addition to all the necessary geodetic instruments, C-band radar transponders to determine whether or not C-band radar tracking stations can track with sufficient accuracy for geodetic work [88].

On 1 January 1972 management responsibility for all DoD geodetic and gravimetric programmes was transferred from the Defense Intelligence Agency to the newly established Defense Mapping Agency [61].

III. The Soviet programme

Although the Soviet Union is known to have an interest in geodesy and mapping, it is difficult to learn from the open sources which of the Cosmos satellites are on geodetic missions. It would be surprising if geodetic satellites were not used by the Soviet Union, since geodetic data are essential for accurate ICBM targeting due to the fact that the gravitational fields around launch and target areas can affect the accuracy with which the re-entry vehicle reaches its target. Since the missile fields in the Soviet Union are spread over a wide geographical area, the use of geodetic satellites becomes almost essential.

It has recently been suggested that some of the satellites thought to be navigational satellites may be on geodetic missions [2]. The satellites have been placed in 1200 to 1400 km orbits with orbital inclinations of about 74° and periods of about 109 and 113 min. Cosmos 800 and 842, which at first sight would appear to belong to the 105-min navigation satellite subset, fly with their orbital planes diametrically opposed to those of the navigation satellite system and may also be geodetic in purpose [89]. Recently, the orbital inclinations of these satellites have been changed to about 83° with orbital periods of about 109 min. The satellites are launched from Plesetsk using an SS-5 (C-1 or Skean IRBM plus an upper stage) vehicle.

The orbital characteristics of these satellites are given in Table 7A.2.

IV. The French programme

In addition to its communications and meteorological satellites, France has also launched a number of geodetic satellites (see Table 7A.3). Its first geodetic satellite was launched in February 1966, using the Diamant A launcher.

162

Appendix 7A

Tables of geodetic satellites

For abbreviations, acronyms and conventions, see page xv.

The designation of each satellite is recognized internationally and is given by the World Warning Agency on behalf of the Committee on Space Research.

More detailed tables of US and Soviet geodetic satellites are to be found in *World Armaments and Disarmament, SIPRI Yearbook 1977* (pp. 152–54, 174); for tables of French geodetic satellites, see p. 176.

Table 7A.1. US geodetic satellites

Satellite name and designation	Launch date and time GMT	Orbital inclination deg	Perigee height km	Apogee height km	Comments
	1958				
NASA Beacon 1 —	23 Oct ..	—	—	—	Failed to orbit
	1959				
NASA Beacon 2 —	14 Aug ..	—	—	—	Failed to orbit
	1962				
USN ANNA 1A —	10 May ..	—	—	—	Failed to orbit
USN ANNA 1B (1962-βμ1)	31 Oct 0810	50	1 077	1 182	
	1964				
USA/USN Secor 1B (1964-01C)	11 Jan 2010	70	904	933	
NASA Beacon —	19 Mar ..	—	—	—	Failed to orbit
NASA Beacon B Explorer 22 (1964-64A)	10 Oct 0307	80	889	1 081	
	1965				
USA/USN/USAF Secor 3 (1965-16E)	9 Mar 1829	70	909	938	

Satellite name and designation	Launch date and time GMT	Orbital inclination deg	Perigee height km	Apogee height km	Comments
USA/USN Secor 2 (1965-17B)	11 Mar 1341	90	296	1 014	
USA/USAF Secor 4 (1965-27B)	3 Apr 2122	90	1 282	1 313	
NASA Beacon C Explorer 27 (1965-32A)	29 Apr 1424	41	941	1 317	
USA/NASA Secor 5 (1965-63B)	10 Aug 1800	69	1 140	2 423	
NASA (Geos 1) Explorer 29 (1965-89A)	6 Nov 1843	59	1 115	2 277	
1966					
USA/USAF Secor 6 (1966-51B)	9 Jun 2010	90	168	3 648	
NASA Pageos 1 (1966-56A)	24 Jun 0014	87	4 207	4 271	A balloon construction, similar to Echo communications satellites
USA/USAF Secor 7 (1966-77B)	19 Aug 1926	90	3 680	3 700	
USAF Secor 8 (1966-89B)	5 Oct 2248	90	3 676	3 706	
1967					
USA/USN Secor 9 (1967-65A)	29 Jun 2107	90	3 803	3 947	
1968					
NASA (Geos 2) Explorer 36 (1968-2A)	11 Jan 1619	106	1 084	1 577	
NASA/USN Secor 10 —	18 May . .	—	—	—	Failed to orbit
USA Secor 11 —	16 Aug . .	—	—	—	Failed to orbit
USA Secor 12 —	16 Aug . .	—	—	—	Failed to orbit
1969					
USA Secor 13 (1969-37B)	14 Apr 0755	100	1 075	1 130	
1970					
NASA/USA Topo 1 (1970-25B)	8 Apr 0824	100	1 064	1 111	

Satellite name and designation	Launch date and time GMT	Orbital inclination deg	Perigee height km	Apogee height km	Comments
	1975				
NASA Geos 3 Explorer 53 (1975-27A)	10 Apr 0000	115	839	853	
	1976				
NASA Lageos (1976-39A)	4 May 0755	110	5 837	5 945	

Table 7A.2. Possible Soviet geodetic satellites

Satellite name and designation	Launch date and time GMT	Orbital inclination deg	Perigee height km	Apogee height km
	1968			
Cosmos 203 (1968-11A)	20 Feb 1605	74	1 178	1 208
Cosmos 256 (1968-106A)	30 Nov 1200	74	1 175	1 227
	1969			
Cosmos 272 (1969-24A)	17 Mar 1648	74	1 181	1 211
Cosmos 312 (1969-103A)	24 Nov 1648	74	1 144	1 179
	1971			
Cosmos 409 (1971-38A)	28 Apr 1438	74	1 177	1 216
Cosmos 457 (1971-99A)	20 Nov 1880	74	1185	1221
	1972			
Cosmos 480 (1972-19A)	25 Mar 0224	83	1 175	1 203
Cosmos 539 (1972-102A)	21 Dec 0155	74	1 343	1 383
	1973			
Cosmos 585 (1973-64A)	8 Sep 0141	74	1 368	1 416
	1974			
Cosmos 650 (1974-28A)	29 Apr 1702	74	1 369	1 402

Satellite name and designation	Launch date and time GMT	Orbital inclination deg	Perigee height km	Apogee height km	Comments
Cosmos 675 (1974-69A)	29 Aug 1453	74	1 365	1 426	
1975					
Cosmos 708 (1975-12A)	12 Feb 0322	69	1 369	1 413	
Cosmos 770 (1975-89A)	24 Sep 1200	83	1 169	1 210	
1976					
Cosmos 807 (1976-22A)	12 Mar 1326	83	396	1 973	

Table 7A.3. French geodetic satellites with possible military applications

Satellite name and designation	Launch date and time GMT	Orbital inclination deg	Perigee height km	Apogee height km
1966				
Diapason 1 (1966-13A)	17 Feb 0838	34	499	2 738
1967				
Diadème 1 (1967-11A)	8 Feb 0930	40	557	1 411
Diadème 2 (1967-14A)	15 Feb 1053	39	591	1 881
1975				
Starlette (1975-10A)	6 Feb 1634	50	807	1 141

8. Interceptor/destructor satellites and FOBSs

All the satellites discussed in the preceding chapters have been potentially dual-purpose; that is, they may be used both for peaceful purposes as well as for waging war. But there also exist two categories of satellites whose function seems to be purely hostile – the interceptor/destructor satellites, and fractional orbital bombardment systems (FOBSs). The development of such interceptor/destructor satellites or, as they are often called, hunter-killer satellites, and systems for destroying, from Earth, satellites in orbit, has opened up space as yet another area in which man can choose to destroy himself. In fact, satellites which hunt down another satellite and destroy it by exploding nearby, have already been tested. This new dimension of the arms race is all the more alarming in that its development has progressed almost totally obscured by the enormous publicity given to the other races to land a man on the Moon or to reach distant planets in the depths of space.

Furthermore, neither the Outer Space Treaty nor the SALT I agreements prohibits the development and testing of anti-satellite systems. The 1967 Outer Space Treaty prohibits the placing or testing of 'weapons of mass destruction' – meaning nuclear weapons only – in space. Similarly, the 1972 US–Soviet SALT I agreements prohibit interference with the other nation's 'national technical means' of verification of compliance with the treaty, meaning reconnaissance satellites, but probably do not include electronic reconnaissance or early-warning satellites, and certainly do not cover communications, navigation, meteorological and geodetic satellites, all of which are also equally important targets for anti-satellite activities.

So far, Earth- and space-based anti-satellite systems have been developed only by the United States and the Soviet Union. In the following sections, various methods of destroying satellites, and the anti-satellite programmes of the two nations are described.

I. Some anti-satellite systems

Basically, two types of concepts have been developed. In one, an anti-satellite system is based on Earth and in the other, a weapon is carried by a satellite in orbit.

The Earth-based anti-satellite systems are of two types – missiles and laser weapons. Missiles carrying nuclear or conventional warheads for intercepting satellites in orbit have been developed and deployed (see page 173, on the US programme). A considerable amount of effort is now being devoted to the development of laser systems, which are briefly discussed below.

Laser is the acronym for Light Amplification by the Stimulated Emission of Radiation. Lasers usually emit light in the infra-red, visible or ultra-violet region of the spectrum of electromagnetic radiation. (See Figure 3.1, page 12) The difference between ordinary light and light from a laser is that the latter is much more intense, directional, monochramatic and coherent. Light from a laser is, therefore, made up of waves which are all of the same wavelength and are all in phase. The size of such a device depends on the material used to produce the laser light, on the power output and on whether the light is emitted in pulses or as a steady beam.

When atoms or molecules of a substance acquire energy from an external source such as an intense light source or electrical energy, they become excited. This excited state is short-lived and the atoms or molecules emit radiation which includes light and return to their initial state. If such a light photon, whose frequency corresponds to the energy difference between the excited state and the initial state, strikes an excited atom, the atom is stimulated and emits a second photon of the same frequency, in phase and in the same direction as the bombarding light photon. The bombarding photon and the emitted photon may then each strike other excited atoms or molecules stimulating further emission of photons. This produces a sudden burst of coherent radiation as all atoms or molecules reach their initial energy state. If the atoms are excited as soon as they reach their initial energy state, a continuous beam of light (laser light) is produced.

Lasers can be broadly classified into three major types. The first type consists of solid-state or 'glass' lasers in which a crystal such as a ruby or yttrium–alumina–garnet is surrounded by an intense flash lamp to excite the atoms and molecules of the crystal. These lasers usually produce intense pulses of light, although production of a continuous beam has been attempted. The second type uses a semiconductor material through which a current is passed. Laser light emitted from such devices is usually in the red or infra-red region and power outputs are generally much smaller than those of other laser types. In the third type, called gas lasers, a gas is excited by means of high voltage applied across a column of gas contained in a tube. These are high-power devices which can be operated to produce a continuous beam of light but the power outputs are not as high as those achieved in pulse solid-state lasers. Many military applications of high-energy lasers (lasers with a power output of one kilowatt and above) have been studied and, in recent years, the development of such lasers against satellites in orbit has gathered some momentum.

It was thought that once high-energy levels were obtained, laser weapons would become an immediate reality. High energies have been produced from electrically excited excimer-type lasers which use noble gases. For example, energies in the single-pulse beam of such a laser have been increased from one joule to greater than 350 joules [90]. Light from such lasers has wavelengths in the visible and ultra-violet regions of the electromagnetic spectrum. However, once high-energy lasers were tested, other problems associated with their use became more apparent. For example, when a high-energy laser beam is used over a great distance, the target has to be very accurately tracked and the beam aimed at it very precisely. Although the short wavelengths involved mean that the optics in an excimer laser can be considerably smaller than those needed in an infra-red laser, the beam damages the optical components of the system since excimer laser pulses carry intense energy. Another serious problem with the use of such systems from the ground against satellites is introduced by the atmosphere, through which the radiation from the laser has to travel before reaching the target.

The fraction of the laser radiation reaching the target is inversely proportional to at least the square of the distance between the point of origin of the laser beam and the target and is very much influenced by the wavelength of the laser beam, particularly in the infra-red spectrum. Laser propagation through the atmosphere also depends on whether the radiation is continuous or pulsed. For example, electric-discharge and chemical lasers can be used in either mode, whereas gas-dynamic lasers are transmitted better if the radiation is continuous. Principal causes of bad propagation are absorption by water vapour, particularly at the Earth's surface and at low altitudes, by carbon dioxide and by aerosol, and scattering due to water vapour and particulate matter.

The beam that reaches its target is no longer the original fine beam but has spread. This is because there is a considerable amount of turbulence in the atmosphere which causes the density to vary greatly. This density variation along the path of the laser beam results in refractive index variations causing the laser beam to spread out. This effect is more pronounced at sea level than at high altitudes. The other factor which causes the beam to become out of focus is called thermal blooming. The air through which the laser beam travels is heated by laser radiation energy. This causes the refractive index of the air to change and under certain circumstances can waste some 90 per cent of the energy of the beam [91]. This effect can be reduced if pulsed lasers are used with short pulse duration so that the air does not have time to heat up enough seriously to deform the laser beam.

Yet another adverse effect for laser transmission through the atmosphere is caused by the electrical breakdown of the air through which the laser beam is travelling. The electrical breakdown of the air generates plasma which absorbs laser energy and effectively shields the target from the laser. The effect is also dependent on the pulse duration so that the energy at which plasma is formed can be considerably increased if the pulse duration is reduced. But this will reduce the total amount of energy delivered to the target.

Table 8.1 illustrates the amount of energy lost from a beam of laser of various wavelengths during its propagation through different atmospheric conditions. The data are for a continuous wave laser with an initial beam diameter of about 50 cm. The target is three kilometres from the laser source and the

Table 8.1. Effects of atmospheric conditions on various lasers

Laser source	Wavelength of emitted radiation m	Amount of energy reaching a target[a] (kW/cm^2)			
		Clear atmosphere		Hazy atmosphere	
		Mild turbulence	Moderate turbulence	Mild turbulence	Moderate turbulence
Carbon dioxide	10.6×10^{-6}	0.09	0.08	0.08	0.06
Carbon monoxide	4.8–6.2 Nominal: 5.0	0.40	0.22	0.30	0.17
Deuterium fluoride	3.7–4.1 Nominal: 3.9	3.00	0.50	2.90	0.37
Hydrogen fluoride	2.5–3.0 Nominal: 2.8	0.15	0.09	0.10	0.07

[a] The target is three kilometres from the laser source. The original beam energy is 25 kW, and the diameter of the beam is about 50 cm.

Source: See reference [91].

initial power of the laser beam, 25 kW. From the table it can be seen that the effect of atmospheric turbulence or clarity is relatively small on radiation from a carbon dioxide laser. On the other hand, although a carbon monoxide laser delivers a greater amount of energy to the target, it is more affected by atmospheric conditions. This effect of atmospheric conditions is greatest on a deuterium fluoride laser but under favourable conditions, the amount of energy received at the target is also greatest – some 30 times the intensity of a carbon dioxide laser under similar conditions.

When a continuous laser beam falls on a target material, heat is produced which generates a cloud of vaporized material and plasma of ionized air. At intensities up to about 10 MW/cm^2 this plasma moves away from the target at subsonic speed [92] and absorbs the beam energy. The effect is to screen the target from the laser beam. At beam intensities of about 100 MW/cm^2, the metallic plasma generates a shock wave which propagates back up the laser beam at supersonic speed. The effect of this is completely to cut off the beam and, therefore, to prevent any thermally induced damage to the target.

Apart from these difficulties, lasers at present tend to be extremely bulky. The power supply for an electric discharge laser, for example, weighs many hundreds of kilograms and occupies several cubic metres in volume. Chemical lasers do not suffer from this problem, but they have relatively poor transmission properties through the atmosphere and substances such as fluorine are highly corrosive and difficult to handle.

With the present state of technology, Earth-based lasers may be ineffective as weapons against orbiting satellites. However, some progress has been made to reduce the effects of turbulence on a laser beam travelling through the atmosphere. The technique being developed is called the coherent optical adaptative technique (COAT) in which controlled distortion of the beam is introduced at the source to compensate for the spread of the laser beam as it travels through the atmosphere [93]. The COAT system can eliminate 70–90 per cent of the distortions caused by atmospheric turbulence [91].

The space-based anti-satellite concepts consist of the use of satellites themselves, lasers carried on board satellites and charged-particle beam weapons. When a satellite itself acts as a weapon, it approaches another satellite in orbit, identifies it and then explodes to destroy or disable the target (see page 175, on the Soviet programme).

As far as space-borne lasers are concerned, these are in the development stage. For such systems, chemical lasers are more useful because of their compact size. Chemical lasers using hydrogen fluoride and built on a small scale are found to have useful efficiencies but effective power intensity has not been achieved. Hydrogen fluoride laser emitting radiation of 2.7×10^{-6} m wavelength can be used in space since the radiation is no longer heavily attenuated and distorted by air. In such systems the problems of precision-aiming at the target, tracking and large optics suitable for high power are still problems to be resolved. Advances in infra-red laser radars show that with such devices, an object in space can be tracked but considerable improvements are still being made to increase the precision with which spacecraft can be tracked in orbit. When such a device becomes available, it can be used to disable solar cells and optical sensors on board a satellite. In high energy lasers and charged-particle beams we may well be seeing the beginning of the next revolution in weapon technology.

A charged-particle beam is a beam of high-energy atomic or sub-atomic particles such as electrons, protons or heavy ions. Such particles are given high energies in huge and very complex apparatus called an accelerator. In all modern high-energy accelerators, an electric field gives energy to the particles and a magnetic field guides them. Two types of accelerator have been developed: linear and circular; the latter are called cyclotrons. In a linear accelerator, the particles are accelerated by a long succession of electric fields and guided in a straight line by a small magnetic field. The circular accelerators – cyclotrons, synchro-cyclotrons and synchrotrons – were developed in order to avoid the use of cumbersome accelerator lengths to produce high-energy beams. The charged particles are accelerated by electric fields, but the magnetic field now guides them in a circular path many times until the required high energies are attained. The highest energy for both electrons and protons has been reached in the synchrotron where the particles are kept moving in an almost circular orbit of fixed radius. The magnet is annular and accelerating fields are provided by one or more radio-frequency devices located at points on the magnetic ring. The frequency of these must be increased at exactly the same rate as the increasing velocity of the particles.

In a conventional accelerator, in which an electric field is used to accelerate a charged particle, sparking in the system sets a limit to the field of about 10 million volts/metre. Another phenomenon, arising from the electric and magnetic fields of the particles themselves and known as the collective effect, also limits the performance of accelerators. At a certain number of particles in the beam, these fields cease to be negligible and can react on the beam in a number of negative ways.

When the collective electric and magnetic fields caused by the acceleration of the particles themselves become comparable to or even stronger than those being generated conventionally and if such fields are controlled, they can be used to enhance the performance of an accelerator. The development of this concept is of particular interest to the present discussion on charged-particle beam weapons. The effect of such a development is that the accelerator can be made more compact and the technological limits mentioned above can be improved upon. The application of the collective effect to accelerate protons was first suggested by a Soviet scientist, Gersh I. Budker as early as 1956. His idea was to inject an intense beam of electrons (equivalent to a current of several thousand amperes) of relatively low energy (only a few MeV) into a weak magnetic field. Such a beam of electrons moving through the magnetic field, radiates electromagnetic energy which results in a shrinking of the beam cross-section. A magnetic field close to the surface of such an intense narrow beam approaches a million gauss which is capable of guiding protons of very high energies.

Recent research on collective effect has also shown considerable promise in improving the electronic field acceleration system. The basic principle is to exploit the large electric fields of particle beams. Two approaches have been considered: one which uses the complex effects of a continuous intense beam of electrons and the other based on very dense clusters of rings of electrons. Experiments using the first method have established that extremely strong electric fields caused by the collective effect – of the order of 50 million volts/metre – exist near the front of the beam which can be used to accelerate ions to useful energies. The exact mechanism of the process is not clearly understood.

A second method of producing intense fields to accelerate ions or particles is also being increasingly investigated. By embedding a proton, for example, in a cluster of electrons which is accelerated, the electrons gain energy of a few MeV, whilst the energy of the proton is some 1836 times that figure. This is because, for a given change in speed, the energy change is directly proportional to the mass of the particle and a proton is 1836 times heavier than an electron. Soviet physicist, Vladimir I. Veksler suggested the use of rings of relativistic electrons rather than a cluster of electrons to accelerate protons and positive ions. In the latter system, the problem is that the electric field which keeps the protons trapped acts on the electrons in such a way as to blow them apart. This effect is absent from a rotating electron ring since the magnetic attraction of the parallel electron currents cancels most of the electrostatic repulsion. Initially a beam of 10- to 20-MeV electrons is curled into a ring by a suitable magnetic field. Such a rotating ring is then introduced into a magnetic solenoid whose axial field becomes weaker with distance. The combination of the rotational motion of the electrons and the small outward component of the magnetic field produces a force which accelerates the ring and protons or

172

positive ions trapped in the ring along the solenoid. In this way the need for the electric accelerating field system is avoided.

There is, however, a problem with such a system. The energy of a proton is not 1836 times that of an electron. This is because, when a ring is accelerated at right angles to its plane, relativistic effects make the rotating electrons appear some 23 times heavier than they are when they are at rest. A proton carried along inside a ring of rotating relativistic electrons will therefore have an energy gain of 1836/23 or about 80. With applied fields of 10 million volts/ metre, proton acceleration rates of 800 MeV/metre could be reached. In practice, however, heavy ions may be used to gain an additional factor in energy.

Research on electron ring accelerators is being carried out not only in the Soviet Union but also in the United States, and some other countries – particularly the People's Republic of China – have recently shown considerable interest. Although there are many problems of a technological nature to be solved before such techniques could be adopted as weapons, the principle of the collective effect ring accelerator has been established.

If such a system is constructed in the future, it will probably be used as an anti-ballistic missile system and will be placed outside the Earth's atmosphere, that is, satellite-based. In this respect, high-energy lasers are more suited to atmospheric propagation. In principle, it is possible to place an accelerator in space particularly when the Space Shuttle system becomes operational. However, the technological problems of providing a large power source for both the accelerator and the radar system used for tracking missiles, and of aiming the particle beam accurately still need to be solved. Aiming is all the more difficult since the charged particles will be affected by the Earth's magnetic field, a problem which a laser beam does not encounter. While the solution of some of these problems may be just a question of time, we may be seeing the spread of the arms race into yet another new dimension.

The US programme

A number of USAF and USN satellite programmes in the early 1960s considered the development of the technology and possibly some hardware for interceptor/ destructor satellite systems. Most of these, however, were either cancelled or suspended, but interest in such systems has recently been revived.

Although in the USA there have been programmes to investigate the possibility of orbital interceptors, initially only an Earth-based anti-satellite system was constructed. The system used missiles based on Johnston and Kwajalein Islands in the Pacific Ocean [94]. In 1963, the USAF demonstrated an anti-satellite capability against a satellite in a low orbit. A Thrust Augmented Thor booster was launched from Johnston Island against a US booster which was orbiting the Earth at the time. The Thor booster came within the calculated kill range of its simulated nuclear warhead [95]. In 1963 President Kennedy referred to the US anti-satellite system which was already under development [96] and the existence of such an operational system was again acknowledged by President Johnson in 1964 [97]. This system was known as Program 437 and Program 922 was intended to produce a follow-on system and investigated the feasibility of a direct-ascent anti-satellite system in which the kill range would be improved and a conventional warhead used. Four direct-ascent anti-satellite vehicles guided by infra-red homing devices were built. Two were launched but neither was successful [95]. However, a ground-launched direct-ascent anti-satellite system is being revived, and will use long-wavelength infra-red seekers for terminal guidance [98].

The US Army also considered a satellite interceptor/destructor system, using Zeus and Nike X missiles, under Program 505 and a system based on these missiles was operational between 1964 and 1968. Another programme, Project 706 or Project SAINT (Satellite Inspector Technique), was conceived in 1960 to demonstrate a military rendezvous with unknown or unidentified spacecraft in Earth orbit. The initial launching of such a system was to have been made in 1962 by an Atlas-Agena B rocket but the project was abandoned before the flight took place [94, 99]. Other USAF programmes related to anti-satellite operations were the PRIOR programme and RMU (Remote Manoeuvring Unit). The latter programme was related to Project SAINT and consisted of investigations of systems to detect, intercept, inspect and destroy hostile satellites. The system would have used television and radio command links.

Some of these programmes were concerned with investigations of inspector satellites. Programme Skipper, an anti-satellite weapon system, was a concept involving vertically launched space mines. This study programme was mainly concerned with satellite destruction systems and was associated with the USN Early Spring programme.

Although there appear to be no US inspector/destructor flights corresponding to Soviet experiments in the field, the Gemini flights (1965–66) successfully demonstrated the US inspector/destructor capability. Gemini 3, launched on 23 March 1965, successfully carried out manned orbital manoeuvres. Gemini 6 and 7 were launched on 15 December and 4 December, respectively. On 15 December a successful rendezvous between these two satellites was carried out in space. Gemini 8, launched on 16 March 1966, docked with the Agena target vehicle on the same day. The Gemini programme was conducted by NASA but the Department of Defense also played a major role in this programme [100]. The latter was interested in the satellite rendezvous technique, particularly with a non-cooperative satellite.

Since 1975, the USAF has been developing a space-based anti-satellite capability similar to that of the Soviet Union. In the US system, a small unarmed homing vehicle is designed to destroy a hostile satellite on impact with the orbiting target. In practice a number of such homing vehicles would be carried into space on board a spacecraft and fired from there against satellites. The homing vehicles would be directed to their prey by a long-wavelength infra-red seeker [95]. The USAF has also shown interest in the development of small ground- and air-launched anti-satellite interceptors, consisting of a nonexplosive interceptor guided by a long-wavelength infra-red homing system to its target satellite, which would be destroyed by collision [101].

The Soviet programme

On 30 October 1967, Cosmos 186 (launched on 27 October using a Soyuz booster) rendezvoused and docked with Cosmos 188 (launched on 30 October). Later, Cosmos 212 did the same with Cosmos 213. While this demonstrated the Soviet capability to manoeuvre satellites to rendezvous with another friendly satellite, a number of experiments have been carried out by the USSR to develop this capability to rendezvous with and destroy an unfriendly satellite. In such experiments, a manoeuvrable satellite was launched to intercept and inspect a target satellite in orbit.

Initially, the interceptor satellite flew close to the target making a high-speed close inspection of it and then moving away before exploding itself. In 1971, however, a new procedure was introduced. The interceptor approached and flew close to the target at a considerably lower speed and in almost the same orbit. After the prolonged inspection, the interceptor was not exploded but was brought down to a lower orbit and allowed to decay. So far no target satellite has actually been destroyed by an interceptor satellite.

In the initial experiments, interceptor/destructor satellites and target satellites have been launched using a rocket system similar to that used for the FOBSs. The rocket system is designated F-1-m where 'm' symbolizes the manoeuvring stage. Until 1971, both these types of satellites were launched from Tyuratam. A new pattern emerged on 9 February 1971 when a target Cosmos 394 satellite was launched from Plesetsk using a C-1 (Skean IRBM plus an upper stage) vehicle. On 25 February the interceptor/destructor Cosmos 397 satellite was launched from Tyuratam using an F-1-m vehicle. Two more such pairs of satellites were launched in 1971 and, after about four years, another such pair of satellites was launched in early February 1976.

It is possible that the satellites launched at the beginning of the interceptor/destructor satellite programme in 1967 may have been to test manoeuvring capability. Cosmos 185 was sent up into a low orbit and was then manoeuvred to a high orbit.

Actual tests of the interceptor/destructor satellite system seem to have started with the launch of Cosmos 217 on 24 April 1968. This satellite was to have been a target satellite but it exploded when it began to make orbital manoeuvres. The 'm' stage may possibly have caused the failure [102]. It was not until six months later, on 19 October 1968, that Cosmos 248 was launched. This delay may well have been caused by the failure of Cosmos 217. Soon after its launch, Cosmos 248 manoeuvred from its lower altitude to an intermediate altitude of about 500 km into a nearly circular orbit. Cosmos 249 was launched on the following day. The satellite, with its 'm' stage, separated from the carrier-rocket, and passed very close to the target satellite, Cosmos 248, as it went into a much higher eccentric orbit. After Cosmos 249 had moved away from Cosmos

248, it was exploded. About a week and a half later, on 1 November, Cosmos 252 was launched, following a mission almost identical to that of Cosmos 249. Another satellite Cosmos 291 was launched on 6 August 1969 but failed to manoeuvre into the 500-km target-satellite circular orbit.

Cosmos 316, launched on 23 December 1969, was orbited at an inclination of 49°, typical for FOBSs. This satellite may belong to the interceptor/destructor satellite system since it manoeuvred more like satellites in this series. A second successful experiment of the interceptor/destructor satellite system involving three satellites was performed on 20–30 October 1970. This pattern was changed, as mentioned above, in 1971 when only two satellites were used to test the interceptor/destructor satellite system. During that year three pairs of satellites were successfully launched. Seven satellites were launched in 1976, and three more by June 1977. Cosmos 909, launched on 19 May, was a target satellite. On 23 May Cosmos 910 was launched but it was reported that the satellite's lifetime was less than one orbit [103]. Cosmos 918, launched on 17 June, may have been a hunter-killer satellite which passed close to Cosmos 909. The second target of the year, Cosmos 959, was launched on 21 October, followed by a hunter-killer, Cosmos 961, launched on 26 October. On 27 October 1977, Japanese newspapers carried reports of observations made by several people of unidentified flying objects. These had been seen on 26 October at 0830 GMT. Several bright objects were seen eventually falling into the Pacific Ocean. Cosmos 961 in fact re-entered the Earth's atmosphere and its predicted position was 36.7°N, 143.7°E off the coast of north-east Japan on 26 October at 0838 GMT. An interesting aspect of these hunter-killer satellites is that each was in orbit for less than a day and that they approached their targets from lower altitudes. This is the new, so-called 'pop-up' technique.

The orbital characteristics of these satellites are given in Table 8A.1.

II. Satellite tracking systems and facilities

In order for interceptor/destructor systems to be effective, it is necessary to know the position of an enemy satellite in space so that an interceptor/destructor satellite can be precisely guided to the target. Tracking systems can be divided into three types depending on the kind of instruments used to observe the satellite being tracked. In one, a satellite is observed by the natural light which is reflected off the satellite. If a satellite transmits a radio beacon or an optical signal, then the detection of this signal forms the basis of the second type of tracking system. In the third system, the satellite is artificially illuminated by a radar or laser beams. In all three types, the satellite has to be visible in some part of the electromagnetic spectrum to the instruments of the ground-based tracking systems. The basic principles employed in these different tracking systems are briefly considered below.

176

Optical tracking of Sun-illuminated satellites can be made using telescopes. In order to track a satellite precisely, after its approximate location has been established, a specially built camera photographs the satellite against the familiar background of the fixed stars. Since the positions of the brighter stars are accurately known, the satellite image on the photograph would give its precise position. The camera used for this purpose is a wide-aperture camera known as the Baker-Nunn camera. With its 30° field of view it can photograph a satellite too faint to be seen with the naked eye. Such a camera is designed for tracking fast-moving satellites and can be used in four different modes. In a stationary camera mode, the tracks of bright satellites are recorded with such a short exposure time that the movement of the fixed stars relative to the Earth is only slight. In a uniform tracking mode, the camera is driven at the same rate as that anticipated for the satellite. With this method brighter images of satellites are obtained so that smaller satellites or those which reflect less sunlight can be tracked. In the oscillating motion mode, two exposures are made – one while tracking the satellite and the other with the camera stationary. Lastly, in the time-exposure mode, the shutter is held open for about 20 s while the camera follows the satellite. This mode is used to help identify the satellite and analyse light variations due to satellite tumbling. The angular accuracy of about two seconds of arc is obtained with a Baker-Nunn camera.

Figure 8.1. Radio signal tracking by measuring satellite beacon signals. The antennas at A and B are separated by distance S. The angle of a satellite is determined by measuring the phase difference ϕ of the radiation arriving at A and B.

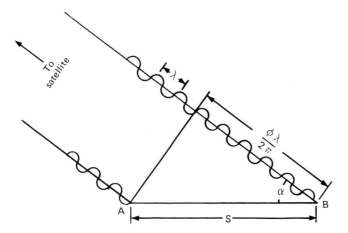

Radio signal tracking by the measurement of satellite beacon signals is simple, accurate and inexpensive but it requires a satellite to carry a beacon transmitter. A plane wavefront of the signal is intercepted by two antennas separated by a known distance S (see Figure 8.1). If a phase difference ϕ is measured between the signals at the two antennas, then the direction of the satellite α is given by

$$\cos \alpha = 2\pi\phi\lambda/S \tag{8.1}$$

177

where λ is the wavelength of the satellite beacon signal. In practice an array of such antennas is constructed and a complete system of antennas at each tracking station forms a cross. A minimum of three separate fixes will provide the six orbital elements.

The accuracy of the measurements will depend on how many coordinated stations are used. However, such a system loses in accuracy when a satellite's orbit is elliptical since a small angular displacement may correspond to large segments of the orbital path. Such a tracking system cannot be used for satellites which do not transmit beacon signals and the method is of little use in tracking synchronous satellites. The system can measure the angular position of a satellite to within 20 minutes of arc (compared with two seconds of arc for a Baker-Nunn system).

Artificial illumination of a satellite with a beam of electromagnetic radiation such as radar or laser will allow the satellite's range and velocity to be determined. Conventional radars direct a narrow pulsed beam at the satellite to be tracked. A small fraction of the energy striking the satellite will be reflected and intercepted by a ground antenna. Measurement of the time taken for the radar signal to reach the satellite and return and the measurement of the Doppler-frequency shifts give the satellite's range and velocity. Radar beams are too broad to allow precise angular measurements but the satellite's range and velocity are adequate to determine its orbit. Tracking satellites using lasers is similar to radar tracking. In practice, the laser beam is so narrow and pencil-like that the location of the satellite has to be known accurately before lasers can be used, which has limited their deployment.

US facilities

Ground support for US military and military-related programmes is mainly derived from the USAF's ETR and Space and Missile Test Center (SAMTEC), the US Army's White Sands Missile Range (WSMR) and Kwajalein Missile Range, and the USN's Pacific Missile Test Center (PMTC). The ETR, apart from providing launch and data acquisition facilities, also has a tracking station which has recently been improved and modernized. SAMTEC manages, operates and maintains the WTR in support of the DoD's space programme providing range tracking as well as data acquisition and other necessary support for all space launches. SAMTEC is also actively engaged in planning for the first space shuttle launch from WTR. The USAF's Space Detection and Tracking System (SPADATS) is primarily a data-processing and -cataloging operation.

The US Army's WSMR, among other things, provides DoD facilities for data acquisition and analysis, and surveillance. The Kwajalein Missile Range has radars for back-up tracking data.

The US Navy's PMTC provides support range services for most launches from the Vandenberg Air Force Base. The DoD's Space Surveillance System (SPASUR), maintained by the Navy, employs radars to illuminate all satellites crossing over the United States. The transmitting stations and stations receiving the reflected radar signals are placed across the continent in an east–west direction from Brown Field, California to Fort Stewart, Florida. A few high-powered transmitters continuously illuminate the sky. The USN also maintains TRANET (Transit Network) to support the Transit Navigation satellite programme.

This section gives only a broad outline of the US tracking facilities. In reality the DoD maintains hundreds of radars, Baker-Nunn cameras, interferometers and other equipment all over the world. The locations and capabilities of many are classified. The information collected by all of these facilities is passed on to NORAD, the North American Air Defense Command.

Soviet facilities

Details of the Soviet satellite tracking systems are scanty. It is, however, reasonable to assume that the Soviet Union has a network of satellite tracking stations probably equipped, as in the case of the United States, with receivers to measure Doppler shifts in radio signals, tracking radars and optical tracking instruments. In 1975, with the Apollo–Soyuz Test Project, locations of some of the tracking stations became known [104]. There are seven Soviet land-based tracking stations: Yevpatoriya, Tbilisi, Dzhusaly, Kolpashevo, Ulan-Ude, Ussuriyisk and Petropavlovsk. It has been suggested that the Soviet equivalent to the US NORAD system may be its ABM defence system [25]. An ABM system is not only a missile launching system but is also a missile tracking system which uses large arrays of radar. A system which is able to track long-range missiles can also track satellites crossing over such systems.

The Soviet Union does not have as extensive a satellite tracking system outside its borders as does the United States. There are some possible land-based stations, for example, two in the United Arab Republic (these may not be operational at present), one in Mali, one in Guinea in West Africa, one on Cuba and one at Fort Lamy, Chad [46]. Because of the limited number of foreign-based tracking stations, the Soviet Union has developed several sea-based satellite tracking stations. These consist of a number of ships operating in the mid-Pacific, in the tropical Atlantic and in the Mediterranean.

III. Fractional orbital bombardment systems (FOBSs)

A fractional orbital bombardment system (FOBS) is designed to place a weapon in orbit and, at a given point, before it has completed its first revolution round the Earth, the weapon is slowed down by a retrorocket and caused to drop on to its target. There are no indications that the USA has either developed or tested such a system but it appears that tests of FOBSs were carried out by the Soviet

Union between 1966 and 1971. The first indication of the development of FOBSs in the Soviet Union came when comments appeared from the USSR that the Scrag SS-10 three-stage carrier-rocket, paraded in Moscow on several occasions from May 1965, was an orbital missile. In November 1967 a large new rocket, Scarp SS-9, was paraded in Moscow and it was reported that the rocket could be used for intercontinental and orbital launching. The SS-9 is bottle-shaped and about 34 m in length. It consists of a large first stage about 3 m in diameter topped by a tapered section and then a smaller portion about 1 m in diameter. When used for space flights, it is possible that four stages are involved, in which case the total length may be about 47 m.

If used as an ICBM, the SS-9 is capable of carrying a 20- to 25-Mt warhead weighing about 4500 kg but its use in FOBSs will reduce the size of the warhead to 10 Mt and its weight to about 3200 kg [105]. The rocket system is called the F-1-r if designed for FOBS, where 'r' symbolizes the retrofire fourth stage which drives the warhead back to Earth, leaving the rest of the system in orbit. An advantage to the Soviet Union of FOBSs for delivering nuclear warheads would be that the US warning time would be reduced. Secondly, US defences could be penetrated from the south, the least defended US front. It must be noted that these tests did not constitute violations of the treaty and resolution banning weapons of mass destruction from orbit, both because they did not complete one orbit and in all likelihood did not carry an actual warhead while undergoing tests, and also because they did not cross over the US mainland.

In 1966, debris was detected in orbit on two occasions from two unannounced flights from the Soviet Union. The debris was located at several altitudes suggesting that explosions may have occurred. The first flight appears to have been made on 17 September 1966 from Tyuratam at an orbital inclination of 49.63° and the second flight occurred on 2 November 1966 with very similar orbital parameters. Since both these launches were unannounced, they are listed in the RAE *Table of Earth Satellites* as Cosmos U.1 and U.2. It was not until 1967 and 1971 that it became apparent that these were the beginning of the FOBSs programme. All such subsequent flights, beginning with Cosmos 139 launched on 25 January 1967, were announced by the Soviet Union under the Cosmos series; their perigees and apogees were given but not their orbital periods. This was presumably because the payload part never completed a full orbit since the retrorocket fired, bringing the payload back to the Earth to strike a target in the USSR.

The variation in launch times of the 1967 FOBSs flights was such that the payloads were fairly consistently recovered at local dusk. This suggests that the tests were probably also used to practise detecting low-orbit missiles [84]. At this particular time of day, conditions are such that spurious echoes appear on a radar screen.

These satellites and their orbital characteristics are given in Table 8A.2.

Appendix 8A

Tables of interceptor/destructor satellites and FOBSs

For abbreviations, acronyms and conventions, see page xv.

The designation of each satellite is recognized internationally and is given by the World Warning Agency on behalf of the Committee on Space Research.

More detailed tables of Soviet interceptor/destructor satellites and FOBSs are to be found in *World Armaments and Disarmament, SIPRI Yearbook 1977* (pp. 171–73).

Table 8A.1. Possible Soviet interceptor/destructor satellites

Satellite name and designation	Launch date and time GMT	Orbital inclination deg	Perigee height km	Apogee height km	Comments
	1967				
Cosmos 185 (1967-104A)	27 Oct 0224	64	518	873	Manoeuvrable; intercept test
	1968				
Cosmos 217 (1968-36A)	24 Apr 1605	62	144	262	Target satellite
Cosmos 248 (1968-90A)	19 Oct 0419	62	475	543	Target satellite
Cosmos 249 (1968-91A)	20 Oct 0405	62	493	2 157	Manoeuvrable; passed near Cosmos 248; exploded
Cosmos 252 (1968-97A)	1 Nov 0029	62	531	2 149	Manoeuvrable; passed near Cosmos 248; exploded
	1969				
Cosmos 291 (1969-66A)	6 Aug 0546	62	147	548	Target satellite
Cosmos 316 (1969-108A)	23 Dec 0922	50	152	1 638	Manoeuvrable; possibly related to FOBS
	1970				
Cosmos 373 (1970-87A)	20 Oct 0546	63	472	544	Target satellite
Cosmos 374 (1970-89A)	23 Oct 0419	63	521	2 141	Manoeuvrable; passed near Cosmos 373; exploded
Cosmos 375 (1970-91A)	30 Oct 0210	63	528	2 098	Manoeuvrable; passed near Cosmos 373; exploded

Satellite name and designation	Launch date and time GMT	Orbital inclination deg	Perigee height km	Apogee height km	Comments
1971					
Cosmos 394 (1971-10A)	9 Feb 1858	66	572	614	Target satellite
Cosmos 397 (1971-15A)	25 Feb 1117	66	574	2 202	Manoeuvrable; passed near Cosmos 394; exploded
Cosmos 400 (1971-20A)	18 Mar 2150	66	983	1 006	Target satellite
Cosmos 404 (1971-27A)	4 Apr 1424	66	802	1 010	Manoeuvrable; passed near Cosmos 400; decayed naturally
Cosmos 459 (1971-102A)	29 Nov 1731	66	224	260	Target satellite
Cosmos 462 (1971-106A)	3 Dec 1312	66	230	1 800	Manoeuvrable; passed near Cosmos 459; exploded
1976					
Cosmos 803 (1976-14A)	12 Feb 1258	66	554	618	Target satellite
Cosmos 804 (1976-15A)	16 Feb 0824	66	556	615	Manoeuvrable; passed near Cosmos 803; probably recovered
Cosmos 814 (1976-34A)	13 Apr 1717	65	118	480	Manoeuvrable; passed near Cosmos 803; probably recovered
Cosmos 839 (1976-67A)	8 Jul 2107	66	984	2 098	Target satellite
Cosmos 843 (1976-71A)	21 Jul 1522	65	132	346	Manoeuvrable; probably failed
Cosmos 880 (1976-120A)	9 Dec 2010	66	560	617	Target satellite
Cosmos 886 (1976-126A)	27 Dec 1243	66	590	2 295	Manoeuvrable; passed near Cosmos 880; exploded

Table 8A.2. Soviet fractional-orbital bombardment systems

Satellite name and designation	Launch date and time GMT	Orbital inclination deg	Perigee height km	Apogee height km
1966				
Cosmos U.1 (1966-88A)	17 Sep 2234	50	163	1 046
Cosmos U.2 (1966-101A)	2 Nov 0043	50	140	855
1967				
Cosmos 139 (1967-05A)	25 Jan 1355	50	144	210
Cosmos 160 (1967-47A)	17 May 1605	50	137	177

Satellite name and designation	Launch date and time *GMT*	Orbital inclination *deg*	Perigee height *km*	Apogee height *km*	Comments
Cosmos 169 (1967-69A)	17 Jul 1648	50	135	200	
Cosmos 170 (1967-74A)	31 Jul 1648	50	121	252	
Cosmos 171 (1967-77A)	8 Aug 1605	50	138	177	
Cosmos 178 (1967-89A)	19 Sep 1453	50	138	258	
Cosmos 179 (1967-91A)	22 Sep 1410	50	139	207	
Cosmos 183 (1967-99A)	18 Oct 1326	50	130	315	
Cosmos 187 (1967-106A)	28 Oct 1312	50	143	301	
1968					
Cosmos 218 (1968-37A)	25 Apr 0043	50	123	162	
Cosmos 244 (1968-82A)	20 Oct 1341	50	134	158	
1969					
Cosmos 298 (1969-77A)	15 Sep 1605	50	127	162	
1970					
Cosmos 354 (1970-56A)	28 Jul 2248	50	134	178	
Cosmos 365 (1970-76A)	25 Sep 1410	50	133	174	
1971					
Cosmos 433 (1971-68A)	8 Aug 2346	50	112	299	

9. Conclusions

The potential peaceful uses of satellites cannot be disputed. They have proved to be of enormous value in a number of different fields, including communications, meteorology, cartography, navigation and, not least, in verifying arms control agreements.

The verification role of satellites was legalized by the SALT I agreements, which embody an acceptance of the fact that national technical means of verification, including satellites, may be used to check compliance with the agreements. Indeed, since ground resolutions as good as 0.15 m are now feasible, there should be no difficulty in observing and identifying such objects as anti-ballistic missile launchers, large radar installations and intercontinental ballistic missiles, as well as surfaced ballistic missile submarines. It would be equally easy to use satellites for guarding against significant concentrations of armed forces – including tanks, heavy artillery and so on.

An obvious limitation of such control methods, however, is that they cannot be used to check qualitative changes in the military arsenals of states, although the development of certain new weapons would be detectable at the testing stage. And even quantitatively, satellite detection is limited by the fact that for example, in certain weapon systems, such as multiple independently targetable re-entry vehicles, the warheads are enclosed within the missile, and are therefore undetectable except by on-site inspection.

But even for verification of a quantitative limitation of arms, an obligation not to use concealment for impeding verification is essential. This has been provided for in the SALT I agreements, and would similarly need to apply in all other disarmament treaties. The development of multi-spectral sensors carried by satellites, which may make it possible in future to detect certain objects under camouflage, by using infra-red sensors to distinguish the object from its surroundings, does not make this obligation of non-concealment any the less important.

Equally, in all such verification operations, a prohibition on interference with satellites would be essential. The concept of verification by satellites could be jeopardized by developments such as those described in Chapter 8, in techniques whereby satellites can be intercepted and destroyed by other satellites.

Although the use of satellites for verification of arms control agreements is one of the so-called peaceful uses which are accepted and allowed today, it is evident from a study of the missions for which most satellites have been launched during the past few years that in fact they are also used for military purposes. (See Table 9.1 for a list of all potential military satellites launched up to end-1976, classified by type of mission.)

It can be seen from the table that, since 1973, the number of photographic reconnaissance satellites launched each year by the USA and the USSR has been steady at about 5 and 35, respectively, the former number being much smaller due solely to the development of the longer-lived Big Bird satellite in the United States: a satellite launched in 1976 was in orbit for 158 days. A similar trend is observable also for Soviet satellites: Cosmos 905, launched in 1977, was recovered after 30 days in orbit – already more than twice as many days as for earlier Soviet satellites.

The other significant development which can be seen from this table is the emergence of China as a satellite power. In the context of arms control verification this is particularly significant, as it opens the way for other countries to take part in the activity of verification which until 1975 was the monopoly of only the two big powers. Apart from this positive aspect of the entry of China into the satellite club, there are negative implications of the increase also in military activities which satellites facilitate. For China, the surveillance of the missile forces of other nations is necessary if its own missiles are to be a credible deterrent.

A study of the launches of satellites during the past few years suggest that they are not being used solely to verify the implementation of the SALT I agreements. Some of the most striking examples of this are, firstly, the use of reconnaissance satellites during the 1973 Middle East War, when the USSR orbited a succession of Cosmos reconnaissance satellites during the periods immediately before and after the war. In 1974 both the USSR and the USA orbited reconnaissance satellites over Cyprus, Greece and Turkey to observe the army coup on Cyprus on 15 July and the subsequent Turkish invasion of Cyprus on 20–22 July. Secondly, and perhaps a very significant use was when a Soviet area-surveillance satellite Cosmos 922 launched from Plesetsk on 30 July 1977 made two passes over the South African Kalahari Desert region. A week after the recovery of this satellite, another one, Cosmos 932, was launched on 27 July. This satellite was a close-look type satellite which manoeuvred so that it made four good passes over the same region in South Africa. The satellite was recovered on 2 August 1977 and on 6 August the Soviet Union informed the United States that South Africa was secretly preparing to detonate an atomic explosion in the Kalahari Desert. The US Big Bird satellite launched on 27 June 1977 also made four passes on 4, 8, 15 and 26 July 1977 over the presumed test site and again on 2, 6 and 13 August. This time the satellite was manoeuvred. If South Africa were about to test a nuclear device, then indeed, the timely detection of the preparations for such a test by reconnaissance satellites has, at least temporarily, halted the test.

This apparently routine use of military reconnaissance satellites to monitor conflict areas is paralleled by the development of civilian satellites such as the Landsat which seem to be potentially suitable for arms control verification. However, the data obtained from military satellites are kept closely guarded secrets while data from, for example, the Landsat satellites are freely distributed, although it has been reported that the US Department of Defense may limit the image resolution of such unclassified photographs so as to prevent them from

Table 9.1. Summary of possible military satellites, by type of mission

Year	Photographic reconnaissance satellites			Electronic reconnaissance satellites		US MIDAS and Vela satellites		Early-warning satellites		Ocean-surveillance satellites		Navigation satellites	
	USA	USSR	China	USA	USSR	MIDAS	Vela	Other	USSR	USA	USSR	USA	USSR
1958													
1959	6											1	
1960	6					2						2	
1961	13					3						3	
1962	26	5		4		1						1	
1963	17	7		7		2	2					3	
1964	24	12		8		2	2					3	
1965	21	17		5			2					4	
1966	23	21		10	0			1				4	
1967	18	22		8	5		2		2		1	3	
1968	16	29		7	7			1	1		1	1	
1969	12	32		6	11	2		1				0	
1970	9	29		7	10	2		3			1	1	1
1971	7	28		3	15			1			2		2
1972	8	30		3	7			2	1		1	1	3
1973	5	35		2	12			2	1		1	1	3
1974	5	28		3	10				1		2	1	4
1975	4	34	1	2	7			2	2		3	1	4
1976	4	34	1	1	9			1	1	4	2	1	8
Total, by country	224	363	2	76	93	10	12	14	9	4	14	31	25
World total, by mission	589			169		10	12	23		18		56	

providing detailed information to the military.

We have seen from the chapters above that whilst producing data which can be of enormous benefit to humanity, satellite technology is like Pandora's box; from it are emerging all kinds of pernicious refinements of the art of war. Satellites are well on the way to being able to navigate lethal weapons to their targets with a high degree of accuracy; they can predict weather conditions to facilitate bombing; they can be used to determine geographical positions with great precision so that no target can remain obscured; and last, but not least, they increasingly enable areas of war to be controlled remotely.

By September 1977, the United States will have spent about $94 thousand million on its military and civilian space programme [94]. About 60 per cent of all US space flights are military-oriented and about one-third of the total sum is spent on military space activities. In the United States, budget requests give detailed requirements for navigation, communications, geodetic, early-warning and weather satellites, whereas no specific item is mentioned for surveillance satellites, presumably because these requests are contained in a classified part of

Communications satellites					Meteorological satellites				Geodetic satellites			FOBSs	Interceptor/ destructor satellites	Yearly total
USA	USSR	NATO	UK	France	USA	USSR	France	UK	USA	USSR	France	USSR	USSR	
1														1
														7
2					2									14
2					1									22
3					4				1					47
4					3	2								45
3	3				3	2			2					62
7	8				6	4			6					80
11	2				6	2			4		1	2		89
17	5				6	4			1		2	9	1	106
11	4				4	2			1	2		2	4	93
5	2		1		3	2			1	2		1	2	83
3	14	1	1		5	6			1		1	2	3	100
5	21	1			2	4	1	1		2		1	6	102
3	24				4	5				2				94
4	33				2	3				1				105
3	24		2	1	4	6				2				96
5	37			1	3	6			1	2	1			116
11	29	1			3	5			1	1			7	124
100	**206**	**3**	**4**	**2**	**61**	**53**	**1**	**1**	**19**	**14**	**5**	**17**	**23**	
	315					**116**				**38**		**17**	**23**	**1 386**

the budget. No comparable information is available from the Soviet space budget but it may not be very different from the US effort; again, military-oriented satellites comprise a similar portion of Soviet space activities.

The extensive use of surveillance and early-warning satellites has become an important part of US strategic doctrine. Satellites for reconnaissance, early warning, communications, navigation and geodesy are becoming effective for accomplishing these tasks. New technology is now available to enhance their performance and, more importantly, to ensure their survivability against enemy attack. With the development of these types of satellites, there seems to be a trend toward developing new doctrines, such as that of flexible response, which would emphasize limited nuclear war-fighting capabilities at various levels. In this new strategy, space technology provides better centralized command and control over military forces via, for example, effective communications satellites. In a limited exchange of nuclear weapons, it is necessary to obtain a damage assessment for making a prompt response; it is evident from the discussion above that reconnaissance satellites provide this capability. It should be noted that precise target information is required in the current counterforce doctrine for fighting limited nuclear war – again a requirement fulfilled by reconnaissance satellites.

The importance of military satellites is further emphasized by the fact that considerable efforts are being made to increase their survivability. This includes research into anti-jamming devices, protection against the effects of nuclear blast and increased surveillance of space by ground- and space-based sensors. Such a detection system can serve as a warning against satellite attacks.

Of the various types of satellites deployed, navigation and geodetic satellites may well revolutionize strategic and tactical warfare. With the aid of such satellites, it will be possible to guide a missile to within a few metres of its target anywhere in the world, thereby acquiring unprecedented accuracy. In fact it is possible that satellites are already being used for real-time mid-course guidance of long-range strategic missiles. Along with the development of these satellites, investigations are under-way into the development of anti-satellite systems. It was reported that the US Army's high-energy laser shot down, from the ground, a remotely piloted vehicle flying at about 480 km/h [106]. How long will it be before a similar test will be performed in outer space? So far there is no indication that either the United States or the Soviet Union has armed its satellites with laser or charged-particle beam weapons, but initial results indicate that in less than 10 years the United Statds will be able to place a laser weapon on its satellites. The US plan is to place a high-energy laser weapon in space using either a Titan-3C rocket or the space shuttle orbiter, in which case the laser device can be assembled in space [107]. However, while the 'death ray' is yet to be born, systems are being worked out and in some cases even tested to destroy satellites in space using conventional explosives.

In conclusion, while it is important to bear in mind the important role satellites play in verifying arms control agreements and in monitoring conflict areas of the world, it must also be kept in mind that these same satellites play an equally great role for the military in these countries. The former role is given much publicity, but what is generally not so well known is the extent to which this new technology is being used to militarize outer space. And, in addition to this use of satellites by the military, the most alarming aspect of advances in space technology is that they are giving rise to new strategic doctrines – doctrines which may well condition man to believe that limited nuclear war can be fought and that any nation can emerge the victor.

References

1. Baker, R. M. L. and Makemson, M. W., *An Introduction to Astrodynamics* (Academic Press, New York, 1960).
2. Bruce, R. W., Satellite orbit sustaining techniques, *American Rocket Society Journal*, Vol. 31 (September 1961), p. 1237.
3. Beard, D. B. and Johnson, F. J., Charge and magnetic field interaction with satellites, *Journal of Geophysical Research*, Vol. 65 (January 1960), p. 1.
4. Corliss, W. R., *Scientific Satellites* (National Aeronautics and Space Administration, Washington, 1967), p. 117.
5. Jenson, J. *et al.*, *Design Guide to Orbital Flight* (McGraw-Hill, New York, 1962).
6. Perry, G. E., private communication.
7. Brown, N., Reconnaissance from space, *World Today*, Vol. 27, No. 2 (February 1971), p. 70.
8. Robinson, C. A. Jr., Soviet treaty violations detected, *Aviation Week & Space Technology*, Vol. 101, No. 16 (21 October 1974), pp. 14–15.
9. Robinson, C. A. Jr., Soviets hiding submarine work, *Aviation Week & Space Technology*, Vol. 101, No. 19 (11 November 1974), pp. 14–16.
10. Robinson, C. A. Jr., U.S. seeks meeting on SALT violations, *Aviation Week & Space Technology*, Vol. 101, No. 21 (25 November 1974), pp. 18–21.
11. Colvocasesses, A. P., Image resolutions for ERTS, Skylab and Gemini/Apollo, *Photogrammatic Engineering* (January 1972), p. 34.
12. Greenwood, T., *Reconnaissance, Surveillance and Arms Control*, Adelphi Paper No. 88 (International Institute for Strategic Studies, London, 1972).
13. Klass, P. J., Military satellites gain vital data, *Aviation Week & Space Technology*, Vol. 91, No. 11 (15 September 1969), pp. 55–61.
14. Stein, K. J., Vidicon system uses three TV sensors, *Aviation Week & Space Technology*, Vol. 91, No. 11 (31 July 1972), pp. 53–57.
15. Klass, P. J., *Secret Sentries in Space* (Random House, New York, 1971).
16. *Aviation Week & Space Technology*, Vol. 98, No. 16 (16 April 1973), p. 9.
17. Johnsen, K., Air Force assigns top priority to 3-segment Satcom Program, *Aviation Week & Space Technology*, Vol. 98, No. 9 (26 February 1973), p. 19.
18. Glushko, V. P., *Rocket Engines of the Gas Dynamics Laboratory–Experimental Design Bureau* (Novosti Press Publishing House, Moscow, 1975).
19. Perry, G. E., The Cosmos Programme, *Flight International*, Vol. 94, No. 3120 (26 December 1968), pp. 1077–79.
20. Perry, G. E., Recoverable Cosmos satellites, *Spaceflight*, Vol. 14 (1972), pp. 183–84.
21. Takenouchi, T., A launch site in the Kizil Kum Desert?, *Kagaku Asahi* (February 1957), pp. 40–48.
22. Perry, G. E., The Soviet Northern Cosmodrome, *Spaceflight*, Vol. 9 (1967), p. 274.
23. Perry, G. E., Cosmos observation, *Flight International*, Vol. 100, No. 3251 (1 July 1971), pp. 29–32.

24. Perry, G. E., Recoverable Cosmos satellites with scientific missions, *Spaceflight*, Vol. 16 (February 1974), p. 69.

25. *Soviet Space Programs, 1971–75*, Staff Report, Committee on Aeronautical and Space Sciences, US Senate, 30 August 1976 (US Government Printing Office, Washington, 1976), Vol. I.
   (*a*) Soviet Space Programs, 1971–75, chapter 6, annex 2, pp. 457–78.
   (*b*) Soviet Space Programs, 1971–75, chapter 6, annex 1, pp. 453–56.

26. *Aviation Week & Space Technology*, Vol. 103, No. 13 (29 September 1975), p. 13.

27. *Aviation Week & Space Technology*, Vol. 103, No. 14 (6 October 1975), p. 11.

28. Perry, G. E., private communication.

29. *Aviation Week & Space Technology*, Vol. 103, No. 24 (15 December 1975), p. 25.

30. *Aviation Week & Space Technology*, Vol. 103, No. 25 (22 December 1975), p. 38.

31. *Aviation Week & Space Technology*, Vol. 103, No. 16 (20 October 1975), p. 9.

32. McGrath, T. and Perry, G. E., private communication.

33. China successfully launches another man-made Earth satellite, *Peking Review*, Vol. 18, No. 49 (5 December 1975), p. 61.

34. China's satellite, *Spaceflight*, Nol. 18, No. 1 (January 1976), p. 22.

35. China's Earth satellite returns, *Peking Review*, No. 51 (17 December 1976), p. 9.

36. *Le Monde* (18 January 1973).

37. Langereux, P., Les armées s'intéressent aux satellites de reconnaissance et de télé-communications, *Air et Cosmos*, No. 579 (31 May 1975), p. 109.

38. Langereux, P., Décision en février pour le satellite Français d'observation?, *Air et Cosmos*, No. 607 (17 January 1976), p. 44.

39. Langereux, P., Décision prochaine pour le satellite Français d'observation, *Air et Cosmos*, No. 613 (28 February 1976), pp. 34–35.

40. Perry, G. E., Cosmos at 74°, *Flight International*, Vol. 102, No. 3325 (30 November 1972), pp. 7889–90.

41. Perry, G. E., private communication.

42. Klass, P. J., Soviets push ocean surveillance, *Aviation Week & Space Technology*, Vol. 99, No. 11 (10 September 1973), pp. 12–13.

43. Taylor, J. W. R. and Monday, D., *Spies in the Sky* (Ian Allan Ltd., Shapperton, Surrey, 1972), p. 118.

44. *Aviation Week & Space Technology*, Vol. 104, No. 21 (24 May 1976), p. 22.

45. *Aviation Week & Space Technology*, Vol. 104, No. 26 (28 June 1976), p. 11.

46. *Aviation Week & Space Technology*, Vol. 106, No. 8 (21 February 1977), p. 8.

47. *Understanding Soviet Naval Development*, Report prepared by the Director of Naval Intelligence and Chief of Information. Released by Office of the Chief of Naval Operations (Department of Navy, Washington, April 1974), p. 28.

48. Soviet ocean surveillance satellite maneuvers, *Defense Space Business Daily* (16 June 1975), p. 254.

49. *Aviation Week & Space Technology*, Vol. 98, No. 25 (18 June 1973), p. 17.
50. Additional warning satellites expected, *Aviation Week & Space Technology*, Vol. 98, No. 20 (14 May 1973), p. 17.
51. USAF launches new experimental satellites, *Aviation Week & Space Technology*, Vol. 102, No. 25 (23 June 1975), p. 12.
52. *Aviation Week & Space Technology*, Vol. 104, No. 4 (26 January 1976), p. 13.
53. *Aviation Week & Space Technology*, Vol. 106, No. 7 (14 February 1977), p. 9.
54. Sheldon, C. S., *Soviet Space Programs, 1966–70*, Staff Report, Committee on Aeronautical and Space Sciences, US Senate, 9 December 1971 (US Government Printing Office, Washington, 1971).
    (*a*) Soviet Space Programs, chapter 9, p. 338.
    (*b*) Soviet Space Programs, chapter 5, p. 168.
55. Sheldon, C. S., *Soviet Space Programs, 1971*, Supplement to Corresponding Report Covering Period 1966–70, Staff Report, Committee on Aeronautical and Space Sciences, US Senate, April 1972 (US Government Printing Office, Washington, 1972).
56. Soviet Big Bird, *Aviation Week & Space Technology*, Vol. 103, No. 2 (17 November 1975), p. 13.
57. Mueller, G. E. *et al.*, Communication satellites—how high?, *Astronautics*, Vol. 6, No. 7 (July 1961), pp. 42–89.
58. Altman, F. J. *et al.*, *Satellite Communications Reference Data Handbook*, Defense Communications Agency, Document No. AD 746–165 (National Technical Information Service, Springfield, July 1972).
59. Lenorovitz, J. M., CIA satellite data link study revealed, *Aviation Week & Space Technology*, Vol. 106, No. 18 (2 May 1977), pp. 25–26.
60. Wall, V. W., Military communication satellites, *Astronautics & Aeronautics*, Vol. 6, No. 4 (April 1968), pp. 52–57.
61. *Hearings before the Committee on Aeronautical and Space Sciences*, US Senate, Committee on NASA Authorization for Fiscal Year 1974, 93rd Congress (US Government Printing Office, Washington, 1973), pp. 1386–87.
62. Shostak, A., Navy telecommunications, past and present, *Naval Research Reviews*, Vol. 28, No. 12 (December 1975), pp. 1–12.
63. Perry, G. E., private communication.
64. Soviets launch initial spacecraft in Statsionar Satcom network, *Aviation Week & Space Technology*, Vol. 104, No. 2 (12 January 1976), p. 42.
65. Three relay satellites set by Russians, *Aviation Week & Space Technology*, Vol. 103, No. 12 (22 September 1975), p. 17.
66. Johnson, K., Soviet plan seven-satellite global communications net, *Aviation Week & Space Technology*, Vol. 103, No. 24 (15 December 1975), pp. 14–16.

67. van Rossum, G., NATO communications satellite launched, *NATO Letter*, Vol. 18, No. 4 (April 1970), pp. 1–6.
68. NATO'S second communications satellite in orbit, *NATO Letter*, Vol. 19, Nos 3–4, March/April 1971, pp. 16–18.
69. 'Sextius', Premiers essais de télécommunications spatiales militaires en 1977, *Air et Cosmos*, No. 580 (7 June 1975), pp. 54–55.
70. Laurila, S. H., *Electronic Surveying and Navigation* (John Wiley, New York, 1976).
71. Guier, W. H. and Wiffenbach, G. C., A satellite Doppler navigation system, *Proceedings of the Institution of Radio Engineers*, Vol. 48, No. 4 (April 1960), pp. 507–16.
72. New space navigation satellite planned, *Aviation Week & Space Technology*, Vol. 101, No. 2 (15 July 1974), pp. 69–70.
73. The militarization of outer space, *Defense Monitor*, Vol. 4, No. 5 (July 1975), p. 4.
74. Nav-Star techniques to be tested by NTS, *Aviation Week & Space Technology*, Vol. 101, No. 3 (22 July 1974), p. 14.
75. Miller, B., Defense Navstar program progressing, *Aviation Week & Space Technology*, Vol. 104, No. 2 (12 January 1976), pp. 45–50.
76. Klass, P. J., Frequency standard orbital tests set, *Aviation Week & Space Technology*, Vol. 105, No. 15 (11 October 1976), pp. 47–51.
77. GAO cites Navstar slippage, *Aviation Week & Space Technology*, Vol. 106, No. 11 (14 March 1977), p. 22.
78. Klass, P. J., DoD weighs Navstar schedule advance, *Aviation Week & Space Technology*, Vol. 101, No. 22 (2 December 1974), pp. 46–49.
79. Perry, G. E. and Wood, C. D., Identification of a navigation satellite system within the Cosmos programme, *Journal of the British Interplanetary Society*, Vol. 29, No. 5 (May 1976), pp. 307–16.
80. USAF admits weather satellites mission, *Aviation Week & Space Technology*, Vol. 98, No. 11 (12 March 1973), p. 18.
81. *Aviation Week & Space Technology*, Vol. 90, No. 4 (27 January 1969), p. 13.
82. Miller, B., USAF, Navy join in weather program, *Aviation Week & Space Technology*, Vol. 99, No. 23 (3 December 1973), pp. 52–55.
83. McLucas, J. L., A new look from USAF's weather satellites, *Air Force Magazine*, Vol. 56, No. 6 (June 1973), pp. 64–67.
84. Perry, G. E., The Cosmos programme, *Flight International*, Vol. 95, No. 3136 (8 May 1969), pp. 773–79.
85. Sheldon, C. S., The Soviet space program, *Air Force Magazine*, Vol. 58, No. 3 (March 1975), pp. 50–56.
86. Covault, C., Soviets plan weather satellite advances, *Aviation Week & Space Technology*, Vol. 105, No. 22 (29 November 1976), pp. 14–15.
87. King-Hele, D., The shape of the Earth, *Science*, Vol. 192, No. 4246 (25 June 1976), pp. 1293–1300.

88. Corliss, W. R., *Putting Satellites to Work* (National Aeronautics and Space Administration, Washington, 1969).

89. Perry, G. E., Wood, C. D. and Wildman, I., private communication.

90. Klass, P. J., Progress made on high-energy lasers, *Aviation Week & Space Technology*, Vol. 106, No. 10 (7 March 1977), pp. 16–17.

91. Klass, P. J., Advanced weaponry research intensifies, *Aviation Week & Space Technology*, Vol. 103, No. 7 (18 August 1975), p. 36.

92. Klass, P. J., Major hurdles for laser weapons cited, *Aviation Week & Space Technology*, Vol. 99, No. 2 (9 July 1973), p. 38.

93. Yaffee, M. L., Lasers investigated for space propulsion, *Aviation Week & Space Technology*, Vol. 102, No. 16 (21 April 1975), pp. 47–54.

94. Sheldon, C. S., United States and Soviet progress in space. Summary data through 1976 and a forward look, Congressional Research Service, Library of Congress, Report No. QB.1C.Gen, 76-32 SP (2 February 1976).

95. Anti-satellite effort decision awaited, *Aviation Week & Space Technology*, Vol. 104, No. 4 (24 January 1977), p. 19.

96. Anti-satellite weapon, *Aviation Week & Space Technology*, Vol. 79, No. 16 (14 October 1963), p. 25.

97. Anti-satellite system, *Aviation Week & Space Technology*, Vol. 81, No. 12 (21 September 1964), p. 21.

98. Killer technology, *Aviation Week & Space Technology*, Vol. 106, No. 13 (28 March 1977), p. 54.

99. McGuire, V., AF hoping for cheap 'SAINT', *Missiles and Rockets*, Vol. 7, No. 20 (14 November 1960), pp. 38–40.

100. Military space flight rages, *Missiles and Rockets*, Vol. 10, No. 26 (25 June 1962), pp. 13–14.

101. *Aviation Week & Space Technology*, Vol. 102, No. 9 (3 March 1975), p. 9.

102. Gibbons, R. F., Soviet military space tests, *Spaceflight*, Vol. 17, No. 889 (August–September 1975), p. 293.

103. Soviets duplicate killer satellite mission, *Aviation Week & Space Technology*, Vol. 106, No. 22 (30 May 1977), p. 15.

104. *Aviation Week & Space Technology*, Vol. 102, No. 18 (5 May 1975), pp. 42–43.

105. Conrad, T. M., Bombs in orbit, *Space Digest*, Vol. 51, No. 2 (February 1968), pp. 66–68.

106. *Aviation Week & Space Technology*, Vol. 105, No. 8 (28 August 1976), p. 9.

107. *Aviation Week & Space Technology*, Vol. 105, No. 17 (25 October 1976), p. 43.

Glossary of terms

Apogee

That point in an orbit of an Earth satellite which is farthest away from the Earth.

Apsides, line of

The line connecting the perigee and apogee in an Earth orbit.

Argument of perigee

The angle or arc, as seen from the focus of an elliptical orbit, from the ascending node to the closest approach of the orbiting body to the focus.

Aries, first point of

The point on the celestial sphere marking the Sun's position at the instant the Earth passes the vernal equinox.

Ascending node

When a satellite passes the Earth's equatorial plane from the south, the node is called an ascending node and from north to south, a descending node.

Atmospheric drag

Decelerating force due to air acting upon a satellite.

Bandwidth

(1) In an antenna, the range of frequencies within which its performance, in respect to some characteristics, conforms to a specified standard; (2) the number of cycles per second between the limits of a frequency band; (3) in information theory, the information-carrying capacity of a communication channel.

Booster

A rocket used to give extra power during lift-off or at another stage of a spacecraft's flight.

Capsule

A container carried on a rocket or spacecraft with the instruments or photographic films intended to be recovered after the flight.

Carrier rocket

A rocket vehicle used to carry an artificial Earth satellite into an Earth orbit after the initial boost.

Channel width	A band of frequencies radiated by a transmitter or which must be handled by a system to carry out a single conversion.
Decay	Various forces acting on a satellite cause it to re-enter the Earth's atmosphere and eventually to burn up (or decay) in the atmosphere.
Doppler shift	The change in frequency with which energy reaches a receiver when the receiver and the energy source are in motion relative to each other.
Eccentricity	The ratio of the distance between the centre and focus of an ellipse to its semi-major axis.
Ecliptic plane	The plane of the Earth's orbit round the Sun.
Ephemeris	A table of predicted positions of bodies in the solar system.
Epoch	A moment in time to which all position measurements are referred.
Equinox	One of the two points, called equinoctial points, where the celestial equatorial plane intersects the Earth's orbital plane. The term is also used to denote the time that the Earth passes through one of these points, marked by the equality in duration of day and night all over the Earth.
Escape stage	A small rocket engine attached to a satellite to provide it with additional thrust for separating it from the booster vehicle.
Gain (of antenna)	A general term used to denote an increase in signal power in transmission from one point to another. An antenna gain is the ratio of the power transmitted along the beam axis to that of an isotropic radiator transmitting the same total power.
Ground track	The projected path traced out by a satellite over the surface of the Earth.
Guidance system	A system which directs the movements of a rocket or a spacecraft with particular reference to the selection of a flight path.

Inclination (orbital)	The angle of inclination of the orbital plane of a satellite to the Earth's equatorial plane.
Ionosphere	The atmospheric shell characterized by high ion density; its base is at about 75 km and it extends to an indefinite height.
Launcher	A rocket vehicle used to orbit a satellite.
Lifetime (of satellite)	The period of time during which instruments on board a satellite remain functioning, or the period from liftoff to decay of the satellite.
Maser	An amplifier using the principle of microwave amplification by stimulated emission of radiation.
Note	The point at which the orbit of a satellite intersects some particular plane, such as an equatorial plane.
Orbit	The path of a satellite under the influence of the Earth's gravitational force; the path is a closed one so that the satellite returns periodically to the same point in the path.
Orbital elements	A set of six parameters defining the orbit of a satellite; these are the right ascension of the ascending node (Ω), the argument of perigee (ω), the orbital inclination (i), the semi-major axis of the orbit (a), the eccentricity of the orbit (e) and the time of perigee passage (T).
Oriented satellite	A satellite is stabilized in such a way that its sensors always point in a particular direction.
Parking orbit	An orbit of a spacecraft around a celestial body, used for assembly of components or to wait for conditions favourable for departure from the orbit.
Payload	Load which a rocket or a spacecraft carries over and above what is necessary for the operation of the vehicle for its flight; for example, camera equipment etc.
Perigee	That point in an orbit of an Earth satellite which is closest to the Earth.

Period (orbital)	The time required for a satellite to go round the Earth once.
Photon	The elementary quantity or quantum of radiant energy.
Precession	The effects whereby, because the Earth is flat at the poles, the orbit of a satellite is caused to turn round the Earth's axis while keeping its orbital inclination constant.
Resolution	Ground resolution is the ground dimension equivalent to one line at the limit of resolution; photographic resolution is defined as the minimum observable spacing between black and white lines in a standard pattern.
Right ascension	The angular distance (usually stated in time, 24 hours = 360 degrees) eastward along the celestial equator, measured from the first point of Aries to the meridian circle passing through any celestial object.
Rocket	A vehicle which is propelled by the recoil force produced when part of its mass is ejected at high velocity and which propels the satellite.
Satellite, artificial	A man-made spacecraft which revolves round a spatial body, such as the Earth.
Semi-major axis	One-half of the longest diameter of an ellipse.
Semi-minor axis	One-half of the shortest diameter of an ellipse.
Space shuttle	The space shuttle system is composed of the orbiter, an external fuel tank which contains the ascent propellant to be used by the orbiter and two solid rocket boosters. The orbiter and solid rocket boosters are reusable.
Sub-satellite	Often a satellite carries one or more smaller satellites, sub-satellites, which are orbited once the parent satellite is in its orbit.
Terminal (ground)	A place or places on the Earth at which a transmitter or a receiver or both form a link between it and a communications satellite.

Tracking system	Systems such as radars or optical telescopes which follow the movements of satellites in Earth orbit.
Trajectory	The dynamical path followed by an object under the influence of gravity and/or other forces. A satellite's orbit is a trajectory which is closed so that the satellite returns periodically to the same point.
Transponder	A combined receiver and transmitter the function of which is to transmit signals automatically when commanded.
Telemetry	Communication over a distance in which coded signals are used.
Upper stage	A second or later stage in a multistage rocket.
Vernal equinox	That point of intersection of the ecliptic and the celestial equator, occupied by the Sun as it changes from south to north declination. Also called first point of Aries.
Vernier control rocket	Small rockets attached to a satellite to make small adjustments in the orbital flight of the satellite.
Vidicon tube	The vidicon tube is a part of the satellite television camera system whose main component is a layer of photoconductive material. The image of the object is formed on this layer and then converted into electrical signals.
Van Allen radiation belt	The zone of high intensity charged particles surrounding the Earth beginning at altitudes of about 1000 km.

Index

For types of satellite see, e.g., Satellite, communications
For national satellite programes, see, e.g., Satellite (China)

ABM system 179
ABM Treaty 26, 46
AFSATCOM 107
Abbreviations, acronyms xv
Aérospatiale–Thomson–CSF 39
Africa 109
Alaska 44
Antarctic 111
Atlantic Ocean 47, 101, 111, 179
Australia 111

BMEWS 44, 46
Baker-Nunn camera 178
Belgium 111
Bomb, nuclear fission products 47
Bonn 111
Bourges, Yvon 39
Brown Field (California) 177, 178

CNES 39
Canada 111
Cape Kennedy 110, 111, 112, 136
Charged-particle beam weapon 171ff
Chicago 21, 22
China 2, 185
Clipper Bow Programme 43
Comsat 107
Cuba 177
Cyprus 17ff, 185

DSCS 103, 106, 107
DTEN 39, 113
Debré, Michel 39
Defense Navigation Satellite Development Program 136
Direction Technique des Constructions et Armes Navales 113
Doppler shift (system) 132–135, 160, 178, 179
Dzhusaly 179

ECCM 40
ECM 40

FLTSATCOM 107
Fairchild Air Force Base 147
Fort Lamy (Chad) 179
Fort Stewart (Florida) 178

Gaulle, Charles de (President) 148
Germany (Federal Republic) 111
Global Positioning System 137
Glossary of terms 199
Greece 17ff, 183
Greenland 44
Guinea (West Africa) 179

Hsinhua News Agency 38

ICBM 36, 44, 45, 105, 108, 137, 162, 179
IDCSP 105, 106
IDSCS 106
IMEWS 46
IRBM, Skean 175
India 15, 17
Indian Ocean 101, 111
International Telecommunications Union (Frequency Registration Board) 109
Italy 111

Johnson, Lyndon B. (President) 173
Johnston Island 173
Jupiter 2

Kamchatka 109
Kapustin Yar 148
Kennedy, John F. (President) 173
Kettering, Grammar School 3, 36, 80, 138
Khavarovsk 149
Kolpashevo 179

Konstantinov, K. I. 33
Kourou (French Guiana) 39
Kwajalein Island 173
Kwajalein Missile Range 178

LASP 32
LES 106, 107, 111
LTTAT 29
Laser 168ff
—, COAT 170, 171
—, effect of atmosphere on 170
Launcher(s) (France), Ariane 39
—, Diamant 162
Launcher(s) (UK), Black Arrow 149
Launcher(s) (USA), Atlas 31, 105
—, Delta 110, 111
—, Scout 136, 148
—, Thor 29, 41, 111, 136, 146
—, Titan 31, 105, 106, 188
Launcher(s) (USSR), Sapwood 36, 108
—, Scarp 179
—, Scrag 179
—, Soyuz 36
—, Vaskhod 36
—, Vostok 149
Le Bourget 39
Loring Air Force Base 147

Mali 179
Mars 2
Mediterranean 179
Mercury 2
Missile(s), Nike 173
—, testing ranges 178
Monterey (California) 147
Moon 2, 5, 10, 97, 167
Moscow 108, 149

NASA 3, 24, 161, 162, 174
NATO 111, 112
NNSS 135, 136
NORAD 178, 179
NTS 136, 137
Netherlands 111
North America 47, 108
Norway 111
Novosibirsk 149

Obninsk 149
Offutt Air Force Base 147
O'Hare International Airport 24, 26
Outer Space Treaty 2, 167

PDM 37, 80
PMTC 178
PTB 47, 48
Pacific Ocean 101, 173, 179
Petropavlovsk 179
Photography, space 20ff
Plesetsk 1, 35–37, 41, 47, 148, 149, 162, 175
Pompidou, Georges (President) 36
Programme, PRIOR 174
—, Skipper 174
Project, Apollo-Soyuz Test 179
—, MIDAS 45
—, SAINT 174
—, Sextius 113

RCA 145, 146, 147
RMU 172
Radar 41, 162, 178, 179
—, OTH 46
—, transponders 162
Radio systems, Loran 131
— —, Omega 131
Royal Aircraft Establishment (UK) 3, 41

SALT I 2, 14, 26, 37, 46, 167, 184, 185
SAMSO 111
SAMTEC 178
SCORE 105
SHAPE Technical Centre 111
SIPRI Yearbook 49, 82, 91, 93, 114, 139, 150, 163, 181
SLBM 47
Satellite (China) 37, 38
Satellite, communications 6, 97ff
—, classification 97
—, Echo 100, 105, 160
—, Molniya 101, 107–109
—, NATO 111, 112
—, Statsionar 109
—, Symphonie 113
—, Tables listing 114–130, 187
—, Tacsat 107

—, Telstar   101
—, transponder   102–105
Satellite, early-warning   44–47
—, Tables listing   93–96, 186
Satellite,   Earth   Resources   Technology
    38, 43
—, Landsat   21, 26, 34, 35, 38, 43, 185
Satellite (France)   39, 113, 149, 162
Satellite, FOBS   167, 175, 179, 180
—, Tables listing   182, 183, 187
Satellite, geodetic   158ff
—, Anna   161
—, Geos   161, 162
—, Pageos   161
—, Tables listing 163–166, 187
Satellite, ground tracks   13, 15, 41, 42
Satellite,   interceptor/destructor   43,   44,
    167ff
—, Gemini   174
—, systems described   167–173
—, Tables listing   181–183, 187
—, tracking facilities   178, 179
—, tracking systems   176–179
Satellite, meteorological   144ff
—, Eole   149
—, Meteor   148, 149
—, Nimbus   144
—, Prospero   149
—, SMS   146, 147
—, Tables listing   150–157, 187
—, Tiros   146, 148
Satellite, navigation   131ff
—, Navstar system   137
—, NTS   137
—, Tables listing   139–143, 186
—, TIMATION   136
—, Transit   135, 136, 138
Satellite, nuclear-explosion detection   47,
    48
 –, Electron   48
—, Vela   48
Satellite, ocean-surveillance   43, 44
—, Table listing   91–92, 186
—, Whitecloud   43
Satellite,   orbital   characteristics   4ff,   40,
    49ff, 82ff, 91ff, 114ff, 139ff, 144, 150ff,
    160, 163ff

—, orbital perturbations   9ff
Satellite, reconnaissance   12ff
Satellite, reconnaissance, electronic   39–42,
    184
— —, Tables listing   82–90
Satellite,   reconnaissance,   photographic
    12ff
— —, Agena   31
— —, Discovererer   26, 27, 29, 31
— —, SAMOS   29, 45
— —, Tables listing   49–81, 186
Satellite (UK)   149
—, Skynet   110
Satellite (USA), Agena   29, 105
—, Anna   161
—, Big Bird   14, 15, 30, 32, 33, 41
—, Close-look   20, 30
—, Echo   11, 100, 105, 160
—, Gemina   172
—, Geos   161, 162
—, MIDAS   46
—, Navstar   137
—, Tiros   146, 148
—, Vela   48
—, Pageos   161
—, SMS   146, 147
Satellite (USSR), Cosmos   3, 14ff, 34–37,
    41, 42, 44, 46–48, 109, 137, 138,
    148, 149, 162, 174ff, 180, 185
—, Electron   48
—, Meteor   148, 149
—, Molniya   46, 101, 107–109
—, Sputnik   36
—, Statsionar   109
Satellite, Teal Ruby   46
Service,   Central   des   Télécommunications
    et d'Informatique   113
Seychelles   111
Shuang-Cheng-Tzu   38
Skylab satellite   24, 25
Skylab space station   21, 23
Sun   5, 10, 11
—, radiation from   11, 101
—, synchronous orbit   17

Tacsatcom   111
TAT   29, 173

TRANET 178
Tampa Bay (Florida) 24
Tbilisi 179
Tsiolkovsky, K. E. 33
Turkey 17ff, 112, 185
Tyuratam 1, 34, 36, 47, 148, 175, 180

UK 110, 111
UK Signals Research and Development Establishment 111
Ulan-Ude 179
United Arab Republic 179
UN General Assembly 3
USA 1, 2, 36, 97, 101, 111, 147, 167, 179, 185, 190
US Air Force 29, 46, 105, 106, 110, 111, 136, 145–147, 173, 174, 178
—, Automated Weather Network 147
—, Global Weather Center 147
US Army 145, 173, 178
US Army Satellite Communications Agency 111
US Defense Intelligence Agency 162
US Defense Mapping Agency 162
US Department of Defense 29, 46, 105, 161, 162, 174, 178, 185

US MacDill Air Force Base, Florida 24, 25
US Naval Research Laboratory 43
US Navy 43, 105, 107, 135, 136, 145, 147, 173, 174, 178
—, Fleet Numerical Weather Center 147
USSR 1, 2, 33ff, 97, 101, 149, 167, 174, 179, 185, 188
Ussuriyisk 179

Van Allen radiation belt 46, 47
Vandenberg Air Force Base (US) 28, 29, 136, 178
Vladivostok 108

WSMR 178
WWMCCS 105
War, Middle East (1973) 185
Weapon System programme 29
World Warning Agency 49, 82, 91, 93, 114, 139, 150, 163, 181

X-rays 47

Yevpatoriya 179